MEMOIRS
of the
American Mathematical Society

Number 951

Center Manifolds
for Semilinear Equations
with Non-dense Domain
and Applications to Hopf Bifurcation
in Age Structured Models

Pierre Magal
Shigui Ruan

November 2009 • Volume 202 • Number 951 (end of volume) • ISSN 0065-9266

American Mathematical Society
Providence, Rhode Island

2000 *Mathematics Subject Classification.* Primary 35K90, 35L10, 92D25.

Library of Congress Cataloging-in-Publication Data

Magal, Pierre.
 Center manifolds for semilinear equations with non-dense domain and applications to Hopf bifurcation in age structured models / Pierre Magal, Shigui Ruan.
   p. cm. — (Memoirs of the American Mathematical Society, ISSN 0065-9266 ; no. 951)
 "Volume 202, number 951 (end of volume)."
 Includes bibliographical references.
 ISBN 978-0-8218-4653-7 (alk. paper)
  1. Evolution equations 2. Cauchy problem. 3. Bifurcation theory 4. Differential equations, Parabolic. I. Ruan, Shigui, 1963- II. Title.

QA377.3.M34 2009
515′.353—dc22
                                                                                   2009029211

Memoirs of the American Mathematical Society

This journal is devoted entirely to research in pure and applied mathematics.

Subscription information. The 2009 subscription begins with volume 197 and consists of six mailings, each containing one or more numbers. Subscription prices for 2009 are US$709 list, US$567 institutional member. A late charge of 10% of the subscription price will be imposed on orders received from nonmembers after January 1 of the subscription year. Subscribers outside the United States and India must pay a postage surcharge of US$65; subscribers in India must pay a postage surcharge of US$95. Expedited delivery to destinations in North America US$57; elsewhere US$160. Each number may be ordered separately; *please specify number* when ordering an individual number. For prices and titles of recently released numbers, see the New Publications sections of the *Notices of the American Mathematical Society*.

Back number information. For back issues see the *AMS Catalog of Publications*.

Subscriptions and orders should be addressed to the American Mathematical Society, P. O. Box 845904, Boston, MA 02284-5904 USA. *All orders must be accompanied by payment.* Other correspondence should be addressed to 201 Charles Street, Providence, RI 02904-2294 USA.

Copying and reprinting. Individual readers of this publication, and nonprofit libraries acting for them, are permitted to make fair use of the material, such as to copy a chapter for use in teaching or research. Permission is granted to quote brief passages from this publication in reviews, provided the customary acknowledgment of the source is given.

Republication, systematic copying, or multiple reproduction of any material in this publication is permitted only under license from the American Mathematical Society. Requests for such permission should be addressed to the Acquisitions Department, American Mathematical Society, 201 Charles Street, Providence, Rhode Island 02904-2294 USA. Requests can also be made by e-mail to reprint-permission@ams.org.

Memoirs of the American Mathematical Society (ISSN 0065-9266) is published bimonthly (each volume consisting usually of more than one number) by the American Mathematical Society at 201 Charles Street, Providence, RI 02904-2294 USA. Periodicals postage paid at Providence, RI. Postmaster: Send address changes to Memoirs, American Mathematical Society, 201 Charles Street, Providence, RI 02904-2294 USA.

© 2009 by the American Mathematical Society. All rights reserved.
Copyright of individual articles may revert to the public domain 28 years after publication. Contact the AMS for copyright status of individual articles.
This publication is indexed in *Science Citation Index*®, *SciSearch*®, *Research Alert*®, *CompuMath Citation Index*®, *Current Contents*®/*Physical, Chemical & Earth Sciences*.
Printed in the United States of America.

∞ The paper used in this book is acid-free and falls within the guidelines established to ensure permanence and durability.
Visit the AMS home page at http://www.ams.org/

10 9 8 7 6 5 4 3 2 1   14 13 12 11 10 09

Contents

Chapter 1.	Introduction	1
Chapter 2.	Integrated Semigroups	5
Chapter 3.	Spectral Decomposition of the State Space	11
Chapter 4.	Center Manifold Theory	21
4.1.	Existence of center manifolds	23
4.2.	Smoothness of center manifolds	30
Chapter 5.	Hopf Bifurcation in Age Structured Models	45
	Acknowledgments	65
Bibliography		67

Abstract

Several types of differential equations, such as delay differential equations, age-structure models in population dynamics, evolution equations with boundary conditions, can be written as semilinear Cauchy problems with an operator which is not densely defined in its domain. The goal of this paper is to develop a center manifold theory for semilinear Cauchy problems with non-dense domain. Using Liapunov-Perron method and following the techniques of Vanderbauwhede et al. in treating infinite dimensional systems, we study the existence and smoothness of center manifolds for semilinear Cauchy problems with non-dense domain. As an application, we use the center manifold theorem to establish a Hopf bifurcation theorem for age structured models.

Received by the editor June 29, 2007.
Article electronically published on July 22, 2009; S 0065-9266(09)00568-7.
2000 *Mathematics Subject Classification*. Primary 35K90, 37L10; Secondary 92D25.
Key words and phrases. Center manifold, semilinear Cauchy problem, non-dense domain, Hopf bifurcation, age structure model.
Research of the second author was partially supported by NSF grants DMS-0412047, DMS-0715772 and NIH grant R01GM083607.

©2009 American Mathematical Society

CHAPTER 1

Introduction

The classical center manifold theory was first established by Pliss [**88**] and Kelley [**65**] and was developed and completed in Carr [**12**], Sijbrand [**95**], Vanderbauwhede [**104**], etc. For the case of a single equilibrium, the center manifold theorem states that if a finite dimensional system has a nonhyperbolic equilibrium, then there exists a center manifold in a neighborhood of the nonhyperbolic equilibrium which is tangent to the generalized eigenspace associated to the corresponding eigenvalues with zero real parts, and the study of the general system near the nonhyperbolic equilibrium reduces to that of an ordinary differential equation restricted on the lower dimensional invariant center manifold. This usually means a considerable reduction of the dimension which leads to simple calculations and a better geometric insight. The center manifold theory has significant applications in studying other problems in dynamical systems, such as bifurcation, stability, perturbation, etc. It has also been used to study various applied problems in biology, engineering, physics, etc. We refer to, for example, Carr [**12**] and Hassard et al. [**52**].

There are two classical methods to prove the existence of center manifolds. The Hadamard (Hadamard [**47**]) method (the graph transformation method) is a geometric approach which bases on the construction of graphs over linearized spaces, see Hirsch et al. [**55**] and Chow et al. [**19, 20**]. The Liapunov-Perron (Liapunov [**71**], Perron [**87**]) method (the variation of constants method) is more analytic in nature, which obtains the manifold as a fixed point of a certain integral equation. The technique originated in Krylov and Bogoliubov [**69**] and was furthered developed by Hale [**48, 49**], see also Ball [**7**], Chow and Lu [**21**], Yi [**112**], etc. The smoothness of center manifolds can be proved by using the contraction mapping in a scale of Banach spaces (Vanderbauwhede and van Gils [**105**]), the Fiber contraction mapping technique (Hirsch et al. [**55**]), the Henry lemma (Henry [**54**], Chow and Lu [**22**]), among other methods (Chow et al. [**18**]). For further results and references on center manifolds, we refer to the monographs of Carr [**12**], Chow and Hale [**16**], Chow et al. [**17**], Sell and You [**94**], Wiggins [**110**], and the survey papers of Bates and Jones [**8**], Vanderbauwhede [**104**] and Vanderbauwhede and Iooss [**106**].

There have been several important extensions of the classical center manifold theory for invariant sets. For higher dimensional invariant sets, it is known that center manifolds exist for an invariant torus with special structure (Chow and Lu [**23**]), for an invariant set consisting of equilibria (Fenichel [**44**]), for some homoclinic orbits (Homburg [**56**], Lin [**72**] and Sandstede [**90**]), for skew-product flows (Chow and Yi [**24**]), for any piece of trajectory of maps (Hirsch et al. [**55**]), and for smooth invariant manifolds and compact invariant sets (Chow et al. [**19, 20**]).

Recently, great attention has been paid to the study of center manifolds in infinite dimensional systems and researchers have developed the center manifold theory for various infinite dimensional systems such as partial differential equations (Bates and Jones [**8**], Da Prato and Lunardi [**30**], Henry [**54**], Scheel [**93**]), semiflows in Banach spaces (Bates et al. [**9**], Chow and Lu [**21**], Gallay [**45**], Scarpellini [**91**], Vanderbauwhede [**103**], Vanderbauwhede and van Gils [**105**]), delay differential equations (Hale [**50**], Hale and Verduyn Lunel [**51**], Diekmann and van Gils [**34, 35**], Diekmann et al. [**36**], Hupkes and Verduyn Lunel [**58**]), infinite dimensional nonautonomous differential equations (Mielke [**81, 82**], Chicone and Latushkin [**15**]), and partial functional differential equations (Lin et al. [**73**], Faria et al. [**43**], Krisztin [**68**], Nguyen and Wu [**83**], Wu [**111**]). Infinite dimensional systems usually do not have some of the nice properties the finite dimensional systems have. For example, the initial value problem may not be well posed, the solutions may not be extended backward, the solutions may not be regular, the domain of operators may not be dense in the state space, etc. Therefore, the center manifold reduction of the infinite dimensional systems plays a very important role in the theory of infinite dimensional systems since it allows us to study ordinary differential equations reduced on the finite dimensional center manifolds. Vanderbauwhede and Iooss [**106**] described some minimal conditions which allow to generalize the approach of Vanderbauwhede [**104**] to infinite dimensional systems.

Let X be a Banach space. Consider the non-homogeneous Cauchy problem

$$(1.1) \qquad \frac{du}{dt} = Au(t) + f(t), \ t \in [0, \tau], \ u(0) = x \in \overline{D(A)},$$

where $A : D(A) \subset X \to X$ is a linear operator, $f \in L^1((0, \tau), X)$. If $\overline{D(A)} = X$, that is, if $D(A)$ is dense in X, the Cauchy problem has been extensively studied (Kato [**63**], Pazy [**85**]). However, there are many examples (see Da Prato and Sinestrari [**31**]) in which the density condition is not satisfied. Indeed, several types of differential equations, such as delay differential equations, age-structure models in population dynamics, some partial differential equations, evolution equations with nonlinear boundary conditions, can be written as semilinear Cauchy problems with an operator which is not densely defined in its domain (see Thieme [**98, 99**], Ezzinbi and Adimy [**42**], Magal and Ruan [**76**]). Da Prato and Sinestrari [**31**] investigated the existence and uniqueness of solutions to the non-homogeneous Cauchy problem (1.1) when the operator has non-dense domain.

In this paper we present a center manifold theory for semilinear Cauchy problems with non-dense domain. Consider the semiflow generated by the semi-linear Cauchy problem

$$\frac{du}{dt} = Au(t) + F(u(t)), \ t \in [0, \tau], \ u(0) = x \in \overline{D(A)},$$

where $F : \overline{D(A)} \to X$ is a continuous map. A very important and useful approach to investigate such non-densely defined problems is to use the integrated semigroup theory, which was first introduced by Arendt [**3, 4**] and further developed by Kellermann and Hieber [**64**], Neubrander [**84**], Thieme [**98, 99**], see also Arendt et al. [**5**] and Magal and Ruan [**76**]. The goal is to show that, combined with the integrated semigroup theory, we can adapt the techniques of Vanderbauwhede [**103, 104**], Vanderbauwhede and Van Gills [**105**] and Vanderbauwhede and Iooss [**106**] to the context of semilinear Cauchy problems with non-dense domain.

As an application, we will apply the center manifold theory for semilinear Cauchy problems with non-dense domain to study Hopf bifurcation in age structure models. Let $u(t, a)$ denote the density of a population at time t with age a. Consider the following age structured model

(1.2)
$$\begin{cases} \dfrac{\partial u(t,a)}{\partial t} + \dfrac{\partial u(t,a)}{\partial a} = -\mu u(t,a), \ a \in (0, +\infty), \\ u(t,0) = \alpha h\left(\int_0^{+\infty} \gamma(a) u(t,a) da\right), \\ u(0,.) = \varphi \in L_+^1\left((0, +\infty); \mathbb{R}\right), \end{cases}$$

where $\mu > 0$ is the mortality rate of the population, the function $h(\cdot)$ describes the fertility of the population, $\alpha \geq 0$ is considered as a bifurcation parameter. Such age structured models are hyperbolic partial differential equations (Hadeler and Dietz [53], Keyfitz and Keyfitz [66]) and have been studied extensively by many researchers since the pioneer work of W. O. Kermack and A. G. McKendrick (Anderson [1], Diekmann et al. [32], Inaba [61]). We refer to some early papers of Gurtin and MacCamy [46] and Webb [107], the monographs by Hoppensteadt [57], Webb [108], Iannelli [59], and Cushing [27], a recent paper of Magal and Ruan [76] and the references therein.

The existence of non-trivial periodic solutions in age structured models has been a very interesting and difficult problem, however, there are very few results (Cushing [25, 26], Prüss [89], Swart [96], Kostava and Li [67], Bertoni [10]). It is believed that such periodic solutions in age structured models are induced by Hopf bifurcation (Castillo-Chavez et al. [13], Inaba [60, 62], Zhang et al. [114]), but there is no general Hopf bifurcation theorem available for age structured models. In this paper we shall use the center manifold theorem for semilinear Cauchy problems with non-dense domain to establish a Hopf bifurcation theorem for the age structured model (1.2).

The paper is organized as follows. In Chapter 2, some results on integrated semigroups are recalled. One of the main tools to develop the center manifold theory is the spectral decomposition of the state space X. The difficulty here is that from the classical theory of C^0-semigroup we only have spectral decomposition of the space $X_0 := \overline{D(A)}$. But in order to deal with non-densely defined problems we need spectral decomposition of the whole state space X. In Chapter 3, we address this issue. In Chapter 4 we present the main results of the paper, namely the existence and smoothness of the center manifold for semilinear Cauchy problems with non-dense domain, by using the Liapunov-Perron method and following the techniques and results of Vanderbauwede and Iooss [106].

In Chapter 5, we apply the center manifold theory to study Hopf bifurcation in the age structured model (1.2). This kind of problems has been considered by Diekmann and van Gils [34, 35] and Diekmann et al. [33] by studying the equivalent integral/delay equations. Nevertheless, here we regard this problem as an example simple enough to illustrate our results. One may observe that the approach used for this kind of problems can be used to study some other types of equations, such as functional differential equations. Once again one of the main difficulties is to obtain the spectral state decomposition for functional differential equations. Notice that this question has been recently addressed for delay differential equations in the space of continuous functions by Liu, Magal and Ruan [74] and for neutral delay differential equations in L^p space by Ducrot, Liu and Magal [39]. Thus, using these recent developments it is also possible to apply our results presented here to functional differential equations. Of course in the context of functional differential equations this problem was considered in the past (see Hale [**50**]). However, the approach presented here allows us to consider both functional differential equations and age-structured problems as special cases of the non-densely defined problem (Magal and Ruan [**76**]).

CHAPTER 2

Integrated Semigroups

In this chapter we recall some results about integrated semigroups. We refer to Arendt [**3, 4**], Neubrander [**84**], Kellermann and Hieber [**64**], Thieme [**99**], and Arendt *et al.* [**5**] for more detailed results on the subject. The results that we present here are taken from Magal and Ruan [**76, 78**].

Let X and Z be two Banach spaces. Denote by $\mathcal{L}(X, Z)$ the space of bounded linear operators from X into Z and by $\mathcal{L}(X)$ the space $\mathcal{L}(X, X)$. Let $A : D(A) \subset X \to X$ be a linear operator. We denote by $R(A)$ the range of A and $N(A)$ the null space of A. If A is the infinitesimal generator of a strongly continuous semigroup of bounded linear operators on X, we denote by $\{T_A(t)\}_{t \geq 0}$ this semigroup. Recall that A is **invertible** if A is a bijection from $D(A)$ into X and A^{-1} is bounded. If X is a $\mathbb{C}$-Banach space, we recall that the **resolvent set of** A is defined by $\rho(A) = \{\lambda \in \mathbb{C} : \lambda I - A \text{ is invertible}\}$. Moreover, we denote by $\sigma(A) := \mathbb{C} \setminus \rho(A)$ the **spectrum of** A.

Note that if X is a real Banach space, then as in Schaefer [**92**, p.134], we can consider the complexification $X^{\mathbb{C}}$ of X, which is the additive group $X \times X$ with scalar multiplication defined by

$$(\alpha, \beta)(x, y) := (\alpha x - \beta y, \beta x + \alpha y)$$

for $(\alpha, \beta) \in \mathbb{C}$ and $(x, y) \in X \times X$. Then $X^{\mathbb{C}}$ is a complex Banach space endowed with the norm

$$\|(x, y)\|_{X^{\mathbb{C}}} = \sup_{0 \leq \theta \leq 2\pi} \|\cos(\theta) x + \sin(\theta) y\|.$$

Define $A^{\mathbb{C}} : D(A^{\mathbb{C}}) \subset X^{\mathbb{C}} \to X^{\mathbb{C}}$ by

$$A^{\mathbb{C}}(u, v) = (Au, Av), \quad \forall (u, v) \in D(A^{\mathbb{C}}) = D(A) \times D(A).$$

Then $A^{\mathbb{C}}$ is a $\mathbb{C}$-linear operator on $X^{\mathbb{C}}$. Set

$$\rho(A) := \rho\left(A^{\mathbb{C}}\right) \text{ and } \sigma(A) := \mathbb{C} \setminus \rho\left(A^{\mathbb{C}}\right).$$

Note that if X is a real Banach space, then it is easy to see that

$$\lambda \in \rho(A) \cap \mathbb{R} \Leftrightarrow \lambda I - A \text{ is invertible}.$$

Let Y be a subspace of X. Y is said to be **invariant** by A if

$$A(D(A) \cap Y) \subset Y.$$

Denote by $A|_Y : D(A|_Y) \subset Y \to X$ the **restriction of** A **to** Y, which is defined by

$$A|_Y x = Ax, \quad \forall x \in D(A|_Y) = D(A) \cap Y.$$

Denote by $A_Y : D(A_Y) \subset Y \to Y$ the **part of** A **in** Y, which is defined by

$$A_Y x = Ax, \quad \forall x \in D(A_Y) = \{x \in D(A) \cap Y : Ax \in Y\}.$$

For convenience, from now on we define
$$X_0 := \overline{D(A)} \text{ and } A_0 := A_{X_0}.$$

LEMMA 2.1. *Let $(X, \|.\|)$ be a $\mathbb{K}$-Banach space (with $\mathbb{K} = \mathbb{R}$ or $\mathbb{C}$) and let $A : D(A) \subset X \to X$ be a linear operator. Assume that $\rho(A) \neq \emptyset$, then*
$$\rho(A_0) = \rho(A).$$

Moreover, we have the following:

(i) *For each $\lambda \in \rho(A_0) \cap \mathbb{K}$ and each $\mu \in (\omega, +\infty)$,*
$$(\lambda I - A)^{-1} = (\mu - \lambda)(\lambda I - A_0)^{-1}(\mu I - A)^{-1} + (\mu I - A)^{-1}.$$

(ii) *For each $\lambda \in \rho(A) \cap \mathbb{K}$,*
$$D(A_0) = (\lambda I - A)^{-1} X_0 \text{ and } (\lambda I - A_0)^{-1} = (\lambda I - A)^{-1}|_{X_0}.$$

PROOF. Without loss of generality we can assume that X is a complex Banach space. Assume that $\lambda \in \rho(A_0)$, $\mu \in \rho(A)$, and set
$$L = (\mu - \lambda)(\lambda I - A_0)^{-1}(\mu I - A)^{-1} + (\mu I - A)^{-1}.$$
Then one can easily check that
$$Lx \in D(A), \ (\lambda I - A) Lx = x, \ \forall x \in X,$$
and
$$L(\lambda I - A) x = x, \forall x \in D(A).$$
Thus, $(\lambda I - A)$ is invertible and $(\lambda I - A)^{-1} = L$ is bounded, so $\lambda \in \rho(A)$. This implies that $\rho(A_0) \subset \rho(A)$. To prove the converse inclusion, we fix $\lambda \in \rho(A)$. Then one can easily prove (ii). So $\rho(A) \subset \rho(A_0)$, and the result follows. □

The following Lemma was proved in Magal and Ruan [**76**, Lemma 2.1].

LEMMA 2.2. *Let $(X, \|.\|)$ be a Banach space and $A : D(A) \subset X \to X$ be a linear operator. Assume that there exists $\omega \in \mathbb{R}$ such that $(\omega, +\infty) \subset \rho(A)$ and*
$$\limsup_{\lambda \to +\infty} \lambda \left\|(\lambda I - A)^{-1}\right\|_{\mathcal{L}(X_0)} < +\infty.$$

Then the following assertions are equivalent:

(i) $\lim_{\lambda \to +\infty} \lambda (\lambda I - A)^{-1} x = x, \forall x \in X_0$.
(ii) $\lim_{\lambda \to +\infty} (\lambda I - A)^{-1} x = 0, \forall x \in X$.
(iii) $\overline{D(A_0)} = X_0$.

Recall that A is a **Hille-Yosida** operator if there exist two constants, $\omega \in \mathbb{R}$ and $M \geq 1$, such that $(\omega, +\infty) \subset \rho(A)$ and
$$\left\|(\lambda I - A)^{-k}\right\|_{\mathcal{L}(X)} \leq \frac{M}{(\lambda - \omega)^k}, \ \forall \lambda > \omega, \ \forall k \geq 1.$$

In the following, we assume that A satisfies some weaker conditions

ASSUMPTION 2.3. *Let $(X, \|.\|)$ be a Banach space and $A : D(A) \subset X \to X$ be a linear operator. Assume that*

(a) There exist two constants, $\omega \in \mathbb{R}$ and $M \geq 1$, such that $(\omega, +\infty) \subset \rho(A)$ and
$$\left\|(\lambda I - A)^{-k}\right\|_{\mathcal{L}(X_0)} \leq \frac{M}{(\lambda - \omega)^k}, \quad \forall \lambda > \omega, \; \forall k \geq 1;$$
(b) $\lim_{\lambda \to +\infty} (\lambda I - A)^{-1} x = 0, \forall x \in X$.

By using Lemma 2.2 and Hille-Yosida theorem (see Pazy [85], Theorem 5.3 on p.20), one obtains the following lemma.

LEMMA 2.4. *Assumption 2.3 is satisfied if and only if there exist two constants, $M \geq 1$ and $\omega \in \mathbb{R}$, such that $(\omega, +\infty) \subset \rho(A)$ and A_0 is the infinitesimal generator of a C_0-semigroup $\{T_{A_0}(t)\}_{t \geq 0}$ on X_0 which satisfies $\|T_{A_0}(t)\|_{\mathcal{L}(X_0)} \leq Me^{\omega t}, \forall t \geq 0$.*

We now define the integrated semigroup generated by A. The notion of the generator for an integrated semigroup is taken from Thieme [99].

DEFINITION 2.5. *Let $(X, \|.\|)$ be a Banach space. A family of bounded linear operators $\{S(t)\}_{t \geq 0}$ on X is called* **an integrated semigroup** *if*
(i) $S(0) = 0$.
(ii) *The map $t \to S(t)x$ is continuous on $[0, +\infty)$ for each $x \in X$.*
(iii) $\forall t, r \geq 0$,
$$S(r)S(t) = \int_0^r (S(\tau + t) - S(\tau)) \, d\tau = S(t)S(r).$$

We say that a linear operator $A : D(A) \subset X \to X$ is the **generator** of an integrated semigroup $\{S(t)\}_{t \geq 0}$ if and only if
$$x \in D(A), \; y = Ax \Leftrightarrow S(t)x - tx = \int_0^t S(s)y \, ds, \; \forall t \geq 0.$$

If A is the generator of an integrated semigroup, we use $\{S_A(t)\}_{t \geq 0}$ to denote this integrated semigroup. The following proposition summarizes some properties of integrated semigroups. Assertion (iv) of the following proposition is well known in the context of integrated semigroup generated by a Hille-Yosida operator. We refer to Magal and Ruan [76, Proposition 2.6] for a proof of this result.

PROPOSITION 2.6. *Let Assumption 2.3 be satisfied. Then A generates a unique integrated semigroup $\{S_A(t)\}_{t \geq 0}$ and for each $x \in X$, each $t \geq 0$, and each $\mu > \omega$, $S_A(t)x$ is given by*

$$(2.1) \quad S_A(t)x = \mu \int_0^t T_{A_0}(s) (\mu I - A)^{-1} x \, ds + (\mu I - A)^{-1} x - T_{A_0}(t) (\mu I - A)^{-1} x.$$

Moreover, we have the following properties:
(i) *For all $t \geq 0$ and all $x \in X$,*
$$\int_0^t S_A(s)x \, ds \in D(A), \quad S_A(t)x = A \int_0^t S_A(s)x \, ds + tx.$$
(ii) *The map $t \to S_A(t)x$ is continuously differentiable if and only if $x \in X_0$ and*
$$\frac{dS_A(t)x}{dt} = T_{A_0}(t)x, \quad \forall t \geq 0, \; \forall x \in X_0.$$
(iii) $T_{A_0}(r)S_A(t) = S_A(t+r) - S_A(r), \; \forall t, r \geq 0$.

(iv) If we assume in addition that A is a Hille-Yosida operator, then we have
$$\|S_A(t) - S_A(s)\|_{\mathcal{L}(X)} \leq M \int_s^t e^{\omega \sigma} d\sigma, \quad \forall t, s \in [0, +\infty) \text{ with } t \geq s.$$

From Proposition 2.6, we also deduce that $S_A(t)$ commutes with $(\lambda I - A)^{-1}$ and
$$S_A(t)x = \int_0^t T_{A_0}(l)x \, dl, \quad \forall t \geq 0, \quad \forall x \in X_0.$$
Hence, $\forall x \in X$, $\forall t \geq 0$, $\forall \mu \in (\omega, +\infty)$,
$$(\mu I - A)^{-1} S_A(t)x = S_A(t) (\mu I - A)^{-1} x = \int_0^t T_{A_0}(s) (\mu I - A)^{-1} x \, ds.$$

Moreover, by using formula (2.1) we know that $\{S_A(t)\}_{t \geq 0}$ is an exponentially bounded integrated semigroup. More precisely, for each $\gamma > \max(0, \omega)$, there exists $M_\gamma > 0$, such that $\|S_A(t)\| \leq M_\gamma e^{\gamma t}$. So by using Proposition 3.10 in Thieme [99], we have for each $\lambda > \max(0, \omega)$ that

$$(2.2) \qquad (\lambda I - A)^{-1} x = \lambda \int_0^{+\infty} e^{-\lambda t} S_A(t) x \, dt.$$

We now consider the non-homogeneous Cauchy problem
$$(2.3) \qquad \frac{du}{dt} = Au(t) + f(t), \, t \in [0, \tau], \quad u(0) = x \in \overline{D(A)}.$$

Assume that f belongs to some appropriated subspace of $L^1((0, \tau), X)$.

DEFINITION 2.7. *A continuous map* $u \in C([0, \tau], X)$ *is called* **an integrated solution** *of* (2.3) *if and only if*
$$(2.4) \qquad \int_0^t u(s) ds \in D(A), \quad \forall t \in [0, \tau],$$
and
$$u(t) = x + A \int_0^t u(s) ds + \int_0^t f(s) ds, \quad \forall t \in [0, \tau].$$

From (2.4) we know that if u is an integrated solution of (2.3) then
$$u(t) \in \overline{D(A)}, \quad \forall t \in [0, \tau].$$

LEMMA 2.8. *Let Assumption 2.3 be satisfied. Then for each* $x \in \overline{D(A)}$ *and each* $f \in L^1((0, \tau), X)$, *(2.3) has at most one integrated solution.*

From now on, for each $\widehat{\tau} > 0$ and each $f \in L^1((0, \widehat{\tau}), X)$, we set
$$(S_A * f)(t) := \int_0^t S_A(t - s) f(s) ds, \forall t \in [0, \widehat{\tau}].$$

Note that from Lemma 2.8 in [76], we know that if $f \in C^1([0, \tau], X)$, then the map $t \to (S_A * f)(t)$ is continuously differentiable on $[0, \tau]$. So the following assumption makes sense.

ASSUMPTION 2.9. *Assume that there exist a real number $\tau^* > 0$ and a non-decreasing map $\delta^* : [0, \tau] \to [0, +\infty)$ such that for each $f \in C^1([0, \tau^*], X)$,*

$$\left\| \frac{d}{dt}(S_A * f)(t) \right\| \leq \delta^*(t) \sup_{s \in [0,t]} \|f(s)\|, \; \forall t \in [0, \tau^*],$$

and

$$\lim_{t \to 0^+} \delta^*(t) = 0.$$

The following theorem was proved in Magal and Ruan [**78**].

THEOREM 2.10. *Let Assumptions 2.3 and 2.9 be satisfied. Then for each $\tau > 0$ and each $f \in C([0, \tau], X)$ the map $t \to (S_A * f)(t)$ is continuously differentiable, $(S_A * f)(t) \in D(A), \forall t \in [0, \tau]$, and if we set $u(t) = \frac{d}{dt}(S_A * f)(t)$, then*

$$u(t) = A \int_0^t u(s)ds + \int_0^t f(s)ds, \; \forall t \in [0, \tau].$$

Moreover, there exists a non-decreasing map $\delta : [0, +\infty) \to [0, +\infty)$, such that $\lim_{t \to 0^+} \delta(t) = 0$ and

$$\|u(t)\| \leq \delta(t) \sup_{s \in [0,t]} \|f(s)\|, \; \forall t \in [0, \tau].$$

Furthermore, for each $\lambda \in (\omega, +\infty)$ we have for each $t \in [0, \tau]$ that

$$(2.5) \quad (\lambda I - A)^{-1} \frac{d}{dt}(S_A * f)(t) = \int_0^t T_{A_0}(t-s)(\lambda I - A)^{-1} f(s)ds.$$

As an immediate consequence of Theorem 2.10 we have the following result.

COROLLARY 2.11. *Let Assumptions 2.3 and 2.9 be satisfied. Then for each $\tau > 0$, each $f \in C([0, \tau], X)$, and each $x \in X_0$, the Cauchy problem (2.3) has a unique integrated solution $u \in C([0, \tau], X_0)$ given by*

$$u(t) = T_{A_0}(t)x + \frac{d}{dt}(S_A * f)(t), \; \forall t \in [0, \tau],$$

and

$$\|u(t)\| \leq Me^{\omega t} \|x\| + \delta(t) \sup_{s \in [0,t]} \|f(s)\|, \; \forall t \in [0, \tau].$$

We now consider a bounded perturbation of A. As an immediate consequence of Proposition 2.16 in Magal and Ruan [**76**], we have the following proposition.

PROPOSITION 2.12. *Let Assumptions 2.3 and 2.9 be satisfied. Let $L \in \mathcal{L}(X_0, X)$ be a bounded linear operator. Then $A + L : D(A) \subset X \to X$ satisfies Assumptions 2.3 and 2.9. More precisely, if we fix $\tau_L > 0$ such that*

$$\delta(\tau_L) \|L\|_{\mathcal{L}(X_0, X)} < 1,$$

and if we denote by $\{S_{A+L}(t)\}_{t \geq 0}$ the integrated semigroup generated by $A + L$, then $\forall f \in C([0, \tau_L], X)$,

$$\left\| \frac{d}{dt}(S_{A+L} * f) \right\| \leq \frac{\delta(t)}{1 - \delta(\tau_L) \|L\|_{\mathcal{L}(X_0, X)}} \sup_{s \in [0,t]} \|f(s)\|, \; \forall t \in [0, \tau_L].$$

From now on, for each $\widehat{\tau} > 0$ and each $f \in C\left([0,\widehat{\tau}], X\right)$, we set

$$(S_A \diamond f)(t) := \frac{d}{dt}(S_A * f)(t), \forall t \in [0,\widehat{\tau}].$$

By using the fact that $(S_A \diamond f)(t) \in X_0, \forall t \in [0,\tau]$ and formula (2.5), we have $\forall t \in [0,\tau]$ that

$$(2.6) \quad (S_A \diamond f)(t) = \lim_{\mu \to +\infty} \int_0^t T_{A_0}(t-l)\mu(\mu I - A)^{-1} f(l) dl, \ \forall f \in Z.$$

This approximation formula was already observed by Thieme [98] in the classical context of integrated semigroups generated by a Hille-Yosida operator. From this approximation formulation, we deduce that for each $t, s \in [0,\tau]$ with $s \leq t$, and $f \in C\left([0,\tau], X\right)$,

$$(2.7) \quad (S_A \diamond f)(t) = T_{A_0}(t-s)(S_A \diamond f)(s) + (S_A \diamond f(s+.))(t-s).$$

To conclude this chapter we state a result proved in Magal and Ruan [78]. This result is one of the main tools to investigate semi-linear problems.

PROPOSITION 2.13. *Let Assumptions 2.3 and 2.9 be satisfied. Then for each $\gamma > \omega$, there exists $C_\gamma > 0$, such that for each $f \in C\left(\mathbb{R}_+, X\right)$ and $t \geq 0$,*

$$e^{-\gamma t} \|(S_A \diamond f)(t)\| \leq C_\gamma \sup_{s \in [0,t]} e^{-\gamma s} \|f(s)\|.$$

More precisely, for each $\varepsilon > 0$, if $\tau_\varepsilon > 0$ is such that $M\delta(\tau_\varepsilon) \leq \varepsilon$, then the above inequality is true with

$$C_\gamma = \frac{2\varepsilon \max\left(1, e^{-\gamma \tau_\varepsilon}\right)}{\left(1 - e^{(\omega-\gamma)\tau_\varepsilon}\right)}, \ \forall \gamma > \omega.$$

CHAPTER 3

Spectral Decomposition of the State Space

The goal of this chapter is to investigate the spectral properties of the linear operator A. Indeed, since A_0 is the infinitesimal generator of a linear C^0-semigroup of X_0, we can apply the standard theory to the linear operator A_0. We will recall some basic important results on the spectral theory for C^0-semigroups. Nevertheless, the classical theory does not apply to A since it is non-densely defined. This question will be mainly addressed in Proposition 3.5. As consequences, we will also derive some results for non-homogeneous non-densely defined problem.

We first investigate the properties of projectors which commute with the resolvents of A_0 and the resolvent of A. Then we will turn to the spectral decomposition of the state spaces X_0 and X. Assume $A : D(A) \subset X \to X$ is a linear operator on a complex Banach X. We start with some basic facts.

LEMMA 3.1. *We have the following:*
  (i) *If Y is invariant by A, then $A|_Y = A_Y$ (i.e. $D(A_Y) = D(A) \cap Y$).*
  (ii) *If $(\lambda I - A)^{-1} Y \subset Y$ for some $\lambda \in \rho(A)$, then*

$$D(A_Y) = (\lambda I - A)^{-1} Y, \ \lambda \in \rho(A_Y) \text{ and } (\lambda I_Y - A_Y)^{-1} = (\lambda I - A)^{-1}|_Y.$$

PROOF. (i) Assume that Y is invariant by A, we have

$$D(A_Y) = \{x \in D(A) \cap Y : Ax \in Y\} = D(A) \cap Y = D(A|_Y),$$

so $A|_Y = A_Y$.

(ii) Assume that $(\lambda I - A)^{-1} Y \subset Y$ for some $\lambda \in \rho(A)$. Then we have

$$\begin{aligned} D(A_Y) &= \{x \in D(A) \cap Y : Ax \in Y\} = \{x \in D(A) \cap Y : (\lambda I - A)x \in Y\} \\ &= (\lambda I - A)^{-1} Y, \end{aligned}$$

and the result follows. $\square$

Let $\Pi : X \to X$ be a bounded linear projector on a Banach space X and let Y be a subspace (closed or not) of X. Then we have the following equivalence

(3.1) $$\Pi(Y) \subset Y \Leftrightarrow \Pi(Y) = Y \cap \Pi(X).$$

LEMMA 3.2. *Let $(X, \|.\|)$ be a Banach space. Let $A : D(A) \subset X \to X$ be a linear operator and let $\Pi : X \to X$ be a bounded linear projector. Assume that*

$$\Pi (\lambda I - A)^{-1} = (\lambda I - A)^{-1} \Pi$$

for some $\lambda \in \rho(A)$. Then we have the following
  (i) $\Pi(D(A)) = D(A) \cap \Pi(X)$ and $\Pi\left(\overline{D(A)}\right) = \overline{D(A)} \cap \Pi(X)$.
  (ii) $A\Pi x = \Pi A x, \forall x \in D(A)$.
  (iii) $A_{\Pi(X)} = A|_{\Pi(X)}$.

11

(iv) $\lambda \in \rho\left(A_{\Pi(X)}\right), D(A_{\Pi(X)}) = (\lambda I - A)^{-1} \Pi(X)$ and $\left(\lambda I - A_{\Pi(X)}\right)^{-1} = (\lambda I - A)^{-1} \mid_{\Pi(X)}$.

(v) $(A \mid_{\Pi(X)})_{\overline{D(A\mid_{\Pi(X)})}} = \left(A_{\overline{D(A)}}\right) \mid_{\Pi(\overline{D(A)})}$.

PROOF. We have
$$\Pi(D(A)) = \Pi(\lambda I - A)^{-1}(X) = (\lambda I - A)^{-1} \Pi(X) \subset D(A).$$

Thus, $\Pi(D(A)) \subset D(A)$. Since Π is bounded, we have $\Pi\left(\overline{D(A)}\right) \subset \overline{D(A)}$. So by using (3.1), we obtain $\Pi(D(A)) = D(A) \cap \Pi(X)$ and $\Pi\left(\overline{D(A)}\right) = \overline{D(A)} \cap \Pi(X)$. This proves (i).

Let $x \in D(A)$ be fixed. Set $y = (\lambda I - A)x$. Then
$$\Pi A x = \Pi A (\lambda I - A)^{-1} y = A(\lambda I - A)^{-1} \Pi y = A \Pi x,$$

which gives (ii). Hence, $\Pi(X)$ is invariant by A, and by using Lemma 3.1, we obtain (iii). Moreover, we have
$$(\lambda I - A)^{-1} \Pi(X) = \Pi(\lambda I - A)^{-1} X \subset \Pi(X).$$

So Lemma 3.1 implies (iv). Finally, we have
$$\begin{aligned} D\left((A\mid_{\Pi(X)})_{\overline{D(A\mid_{\Pi(X)})}}\right) &= \left\{x \in D(A\mid_{\Pi(X)}) : Ax \in \overline{D(A\mid_{\Pi(X)})}\right\} \\ &= \left\{x \in \Pi(X) \cap D(A) : Ax \in \overline{D(A)} \cap \Pi(X)\right\} \\ &= \left\{x \in \Pi\left(\overline{D(A)}\right) \cap D(A) : Ax \in \Pi\left(\overline{D(A)}\right)\right\} \\ &= D\left(\left(A_{\overline{D(A)}}\right)\mid_{\Pi(\overline{D(A)})}\right). \end{aligned}$$

This shows that (v) holds. □

LEMMA 3.3. *Let the assumptions of Lemma 3.2 be satisfied. Assume in addition that Π has a finite rank. Then $\Pi(D(A))$ is closed, $\Pi\left(\overline{D(A)}\right) = \Pi(D(A)) \subset D(A)$, and $A \mid_{\Pi(X)}$ is a bounded linear operator from $\Pi(D(A))$ into $\Pi(X)$.*

PROOF. By using Lemma 3.2, we have $\Pi(D(A)) = D(A) \cap \Pi(X)$, so $\Pi(D(A))$ is a finite dimensional subspace of X. It follows that $\Pi(D(A))$ is closed and $A\mid_{\Pi(X)}$ is bounded. Now since Π is bounded, we have $\Pi\left(\overline{D(A)}\right) \subset \overline{\Pi(D(A))} = \Pi(D(A))$, and the result follows. □

LEMMA 3.4. *Let Assumption 2.3 be satisfied. Let $\Pi_0 : X_0 \to X_0$ be a bounded linear projector. Then*

(3.2) $$\Pi_0 T_{A_0}(t) = T_{A_0}(t) \Pi_0, \quad \forall t \geq 0$$

if and only if

(3.3) $$\Pi_0 (\lambda I - A_0)^{-1} = (\lambda I - A_0)^{-1} \Pi_0, \forall \lambda > \omega.$$

If we assume in addition that (3.2) is satisfied, then we have the following:

(i) $\Pi_0 (D(A_0)) = D(A_0) \cap \Pi_0(X_0)$ and $A_0 \Pi_0 x = \Pi_0 A_0 x, \forall x \in D(A_0)$.
(ii) $A_0 \mid_{\Pi(X_0)} = (A_0)_{\Pi_0(X_0)}$.
(iii) $T_{A_0\mid_{\Pi_0(X_0)}}(t) = T_{A_0}(t) \mid_{\Pi_0(X_0)}, \forall t \geq 0$.

(iv) *If we assume in addition that Π_0 has a finite rank, then $\Pi_0(X_0) = \Pi_0(D(A_0)) \subset D(A_0)$, $A_0|_{\Pi_0(X_0)}$ is a bounded linear operator from $\Pi_0(X_0)$ into itself, and*
$$T_{A_0|_{\Pi_0(X_0)}}(t) = e^{A_0|_{\Pi_0(X_0)}t}, \forall t \geq 0.$$

PROOF. $(3.2) \Rightarrow (3.3)$ follows from the following formula
$$(\lambda I - A_0)^{-1} x = \int_0^{+\infty} e^{-\lambda s} T_{A_0}(s) x \, ds, \forall \lambda > \omega, \forall x \in Y.$$

$(3.3) \Rightarrow (3.2)$ follows from the exponential formula (see Pazy [**85**, Theorem 8.3, p.33])
$$T_{A_0}(t)x = \lim_{n \to +\infty} \left(I - \frac{t}{n} A_0\right)^{-n} x, \quad \forall x \in X_0.$$

By applying Lemma 3.2 and Lemma 3.3 to A_0, we obtain (i)-(iv). □

The idea of proving the following result comes from the proof of Theorem 2.6 in Thieme [**102**].

PROPOSITION 3.5. *Let Assumption 2.3 be satisfied. Let $\Pi_0 : X_0 \to X_0$ be a bounded linear projector satisfying the following properties*
$$\Pi_0 (\lambda I - A_0)^{-1} = (\lambda I - A_0)^{-1} \Pi_0, \quad \forall \lambda > \omega$$

and
$$\Pi_0(X_0) \subset D(A_0) \text{ and } A_0|_{\Pi_0(X_0)} \text{ is bounded.}$$

Then there exists a unique bounded linear projector Π on X satisfying the following properties:

(i) $\Pi|_{X_0} = \Pi_0$.
(ii) $\Pi(X) \subset X_0$.
(iii) $\Pi(\lambda I - A)^{-1} = (\lambda I - A)^{-1} \Pi, \forall \lambda > \omega$.

Moreover, for each $x \in X$ we have the following approximation formula
$$\Pi x = \lim_{\lambda \to +\infty} \Pi_0 \lambda (\lambda I - A)^{-1} x = \lim_{h \to 0^+} \frac{1}{h} \Pi_0 S_A(h) x.$$

PROOF. Assume first that there exists a bounded linear projector Π on X satisfying (i)-(iii). Let $x \in X$ be fixed. Then from (ii) we have $\Pi x \in X_0$, so
$$\Pi x = \lim_{\lambda \to +\infty} \lambda (\lambda I - A)^{-1} \Pi x.$$

Using (i) and (iii), we deduce that
$$\Pi x = \lim_{\lambda \to +\infty} \Pi_0 \lambda (\lambda I - A)^{-1} x.$$

Thus, there exists at most one bounded linear projector Π satisfying (i)-(iii).

It remains to prove the existence of such an operator Π. To simplify the notation, set $B = A_0|_{\Pi_0(X_0)}$. Then by assumption, B is a bounded linear operator from $\Pi_0(X_0)$ into itself, and
$$T_{A_0}(t) \Pi_0 x = e^{Bt} \Pi_0 x, \forall t \geq 0, \forall x \in X_0.$$

Let $x \in X$ be fixed. Since $S_A(t)x \in X_0$ for each $t \geq 0$, we have for each $h > 0$ and each $\lambda > \omega$ that
$$(\lambda I - A_0)^{-1} S_A(h) x = S_A(h) (\lambda I - A)^{-1} x = \int_0^h T_{A_0}(h-s)(\lambda I - A)^{-1} x \, ds$$

and
$$\Pi_0 (\lambda I - A_0)^{-1} S_A(h)x = (\lambda I - A_0)^{-1} \Pi_0 S_A(h)x$$
$$= \int_0^h \Pi_0 T_{A_0}(h-s)(\lambda I - A)^{-1} x \, ds$$
$$= \int_0^h e^{B(h-s)} \Pi_0 (\lambda I - A)^{-1} x \, ds.$$

Since B is a bounded linear operator, $t \to e^{Bt}$ is operator norm continuous and
$$\frac{1}{h} \int_0^h e^{B(h-s)} ds = I_{\Pi_0(X_0)} + \frac{1}{h} \int_0^h \left[e^{B(h-s)} - I_{\Pi_0(X_0)} \right] ds.$$

Thus, there exists $h_0 > 0$, such that for each $h \in [0, h_0]$,
$$\left\| \frac{1}{h} \int_0^h \left[e^{B(h-s)} - I_{\Pi_0(X_0)} \right] ds \right\|_{\mathcal{L}(\Pi_0(X_0))} < 1.$$

It follows that for each $h \in [0, h_0]$, the linear operator $\frac{1}{h} \int_0^h e^{B(h-s)} ds$ is invertible from $\Pi_0(X_0)$ into itself and
$$\left(\frac{1}{h} \int_0^h e^{B(h-s)} ds \right)^{-1} = \left(I_{\Pi_0(X_0)} - \left(I_{\Pi_0(X_0)} - \frac{1}{h} \int_0^h e^{B(h-s)} ds \right) \right)^{-1}$$
$$= \sum_{k=0}^{\infty} \left(I_{\Pi_0(X_0)} - \frac{1}{h} \int_0^h e^{B(h-s)} ds \right)^k.$$

We have for each $\lambda > \omega$ and each $h \in (0, h_0]$ that
$$\left(\frac{1}{h} \int_0^h e^{B(h-s)} ds \right)^{-1} (\lambda I - A_0)^{-1} \Pi_0 \frac{1}{h} S_A(h)x = \Pi_0 (\lambda I - A)^{-1} x.$$

Since for each $t \geq 0$, $e^{Bt}\Pi_0 = T_{A_0}(t)\Pi_0$ commutes with $(\lambda I - A_0)^{-1}$, it follows that for each $h \in [0, h_0]$, $\left(\frac{1}{h} \int_0^h e^{B(h-s)} ds \right)^{-1} \Pi_0$ commutes with $(\lambda I - A_0)^{-1}$. Therefore, we obtain for each $\lambda > \omega$ and each $h \in (0, h_0]$ that

$$(3.4) \quad \lambda(\lambda I - A_0)^{-1} \left(\frac{1}{h} \int_0^h e^{B(h-s)} ds \right)^{-1} \Pi_0 \frac{1}{h} S_A(h)x = \Pi_0 \lambda (\lambda I - A)^{-1} x.$$

Now it is clear that the left hand side of (3.4) converges as $\lambda \to +\infty$. So we can define $\Pi : X \to X$ for each $x \in X$ by

$$(3.5) \quad \Pi x = \lim_{\lambda \to +\infty} \Pi_0 \lambda (\lambda I - A)^{-1} x.$$

Moreover, for each $h \in (0, h_0]$ and each $x \in X$,

$$(3.6) \quad \Pi x = \left(\frac{1}{h} \int_0^h e^{B(h-s)} ds \right)^{-1} \Pi_0 \frac{1}{h} S_A(h)x.$$

It follows from (3.6) that $\Pi : X \to X$ is a bounded linear operator and $\Pi(X) \subset X_0$. Furthermore, by using (3.5), we know that $\Pi |_{X_0} = \Pi_0$ and Π commutes with the

resolvent of A. Also notice that for each $h \in (0, h_0]$,
$$\frac{1}{h}\Pi_0 S_A(h)x = \frac{1}{h}\int_0^h e^{B(h-s)}\Pi x ds.$$
So
$$\Pi x = \lim_{h \searrow 0} \frac{1}{h}\Pi_0 S_A(h)x.$$
Finally, for each $x \in X$,
$$\begin{aligned}\Pi\Pi x &= \lim_{\lambda \to +\infty} \Pi\Pi_0 \lambda (\lambda I - A)^{-1} x = \lim_{\lambda \to +\infty} \Pi_0^2 \lambda (\lambda I - A)^{-1} x \\ &= \lim_{\lambda \to +\infty} \Pi_0 \lambda (\lambda I - A)^{-1} x = \Pi x.\end{aligned}$$
This implies that Π is a projector. $\square$

Note that if the linear operator Π_0 has a finite rank, then $A_0 \mid_{\Pi_0(X_0)}$ is bounded. So we can apply the above proposition.

By Proposition 2.6, Lemmas 3.2 and 3.4, we obtain the following results.

LEMMA 3.6. *Let Assumption 2.3 be satisfied. Let $\Pi : X \to X$ be a bounded linear projector. Assume that*
$$\Pi(\lambda I - A)^{-1} = (\lambda I - A)^{-1}\Pi, \quad \forall \lambda \in (\omega, +\infty).$$
Then $A \mid_{\Pi(X)} = A_{\Pi(X)}$ satisfies Assumption 2.3 on $\Pi(X)$. Moreover,

(i) $(A \mid_{\Pi(X)})\overline{D(A\mid_{\Pi(X)})} = \left(A_{\overline{D(A)}}\right)\mid_{\Pi(\overline{D(A)})} = A_0 \mid_{\Pi(X_0)}.$
(ii) $S_A(t)\Pi = \Pi S_A(t), \forall t \geq 0.$
(iii) $S_{A\mid_{\Pi(X)}}(t) = S_A(t) \mid_{\Pi(X)}, \forall t \geq 0.$

From the above results, we obtain the second main result of this chapter.

PROPOSITION 3.7. *Let Assumptions 2.3 and 2.9 be satisfied. Let $\Pi : X \to X$ be a bounded linear projector. Assume that*
$$\Pi(\lambda I - A)^{-1} = (\lambda I - A)^{-1}\Pi, \forall \lambda \in (\omega, +\infty).$$
Then the linear operator $A \mid_{\Pi(X)} = A_{\Pi(X)}$ satisfies Assumptions 2.3 and 2.9 in $\Pi(X)$. Moreover, for each $\tau > 0$, each $f \in C([0, \tau], X)$, and each $x \in X_0$, if we set for each $t \in [0, \tau]$ that
$$u(t) = T_{A_0}(t)x + \frac{d}{dt}(S_A * f)(t),$$
then
$$\Pi u(t) = T_{A_0\mid_{\Pi(X_0)}}(t)\Pi x + \frac{d}{dt}\left(S_{A\mid_{\Pi(X)}} * \Pi f\right)(t),$$
$$\Pi u(t) = \Pi x + A \mid_{\Pi(X)} \int_0^t \Pi u(s)ds + \int_0^t \Pi f(s)ds,$$
and
$$\|\Pi u(t)\| \leq Me^{\omega t}\|\Pi x\| + \delta(t) \sup_{s \in [0,t]} \|\Pi f(s)\|, \forall t \in [0, \tau].$$
Furthermore, if Π has a finite rank and $\Pi(X) \subset X_0$, then $\Pi(X) = \Pi(X_0) \subset \Pi(D(A_0)) \subset D(A_0)$, $A \mid_{\Pi(X)}$ is a bounded linear operator from $\Pi(X_0)$ into itself.

In particular, $A \mid_{\Pi(X)} = A_0 \mid_{\Pi(X_0)}$ and the map $t \to \Pi u(t)$ is a solution of the following ordinary differential equation in $\Pi(X_0)$:

$$\frac{d\Pi u(t)}{dt} = A_0 \mid_{\Pi(X_0)} \Pi u(t) + \Pi f(t), \quad \forall t \in [0, \tau], \quad \text{with } \Pi u(0) = \Pi x.$$

We now recall some well known results about spectral theory of closed linear operators. We first recall that if $\widehat{\lambda} \in \rho(A)$,

$$(3.7) \qquad (\lambda I - A)^{-1} = \left(\widehat{\lambda} I - A\right)^{-1} \sum_{n=0}^{\infty} \left(\widehat{\lambda} - \lambda\right)^n \left(\widehat{\lambda} I - A\right)^{-n},$$

whenever $\left|\lambda - \widehat{\lambda}\right| \left\|\left(\widehat{\lambda} I - A\right)^{-1}\right\|_{\mathcal{L}(X)} < 1$. So one obtains that $(\lambda I - A)^{-1}$ is holomorphic on $\rho(A)$.

The following result is proved in Yosida [**113**, Theorems 1 and 2, p.228-299].

THEOREM 3.8. *Let $A : D(A) \subset X \to X$ be a closed linear operator in the complex Banach space X and let λ_0 be an isolated point of $\sigma(A)$. Then,*

$$(3.8) \qquad (\lambda I - A)^{-1} = \sum_{k=-\infty}^{\infty} (\lambda - \lambda_0)^k B_k,$$

where for each integer k,

$$(3.9) \qquad B_k = \frac{1}{2\pi i} \int_{S_\mathbb{C}(\lambda_0,\varepsilon)^+} (\lambda - \lambda_0)^{-k-1} (\lambda I - A)^{-1} d\lambda,$$

where $S_\mathbb{C}(\lambda_0, \varepsilon)^+$ is the counter-clockwise oriented circumference $|\lambda - \lambda_0| = \varepsilon$ for sufficiently small $\varepsilon > 0$ such that $|\lambda - \lambda_0| \leq \varepsilon$ does not contain other point of the spectrum than λ_0. We have the following properties

$$(3.10) \qquad \begin{array}{l} B_k B_m = 0, \ k \geq 0, m \leq -1, \\ B_n = (-1)^n B_0^{n+1}, \ n \geq 1, \\ B_{-p-q+1} = B_{-p} B_{-q} (p, q \geq 1), \\ B_n = (A - \lambda_0 I) B_{n+1} (n \geq 0), \\ (A - \lambda_0 I) B_{-n} = B_{-(n+1)} = (A - \lambda_0 I)^n B_{-1}, \\ (A - \lambda_0 I) B_0 = B_{-1} - I. \end{array}$$

Note that from the third equation of (3.10), we have for each $p \geq 1$ that

$$B_{-p} B_{-1} = B_{-p-1+1} = B_{-p},$$

so B_{-1} is a projector on X. Since

$$(A - \lambda_0 I) B_{-1} = B_{-2},$$

it follows that

$$AB_{-1} = \lambda_0 B_{-1} + B_{-2}.$$

So A restricted to $R(B_{-1})$ is a bounded linear operator. We also have for each $p \geq 1$ that

$$(3.11) \qquad AB_{-p} = AB_{-1}B_{-p} = \lambda_0 B_{-1} B_{-p} + B_{-2} B_{-p} = \lambda_0 B_{-p} + B_{-p-1}.$$

Moreover, from (3.9) it is clear that B_{-1} commutes with $(\lambda I - A)^{-1}$ for each $\lambda \in \rho(A)$. Thus,

$$\left(\lambda_0 I - A \mid_{B_{-1}(X)}\right)^{-1} = (\lambda_0 I - A)^{-1} \mid_{B_{-1}(X)}.$$

Furthermore, by using the last equation of (3.10), we deduce that $\lambda_0 \notin \sigma\left(A\mid_{(I-B_{-1})(X)}\right)$ and
$$\left(\lambda_0 I - A\mid_{(I-B_{-1})(X)}\right)^{-1} = B_0\mid_{(I-B_{-1})(X)}.$$
Recall that λ_0 **is a pole of** $(\lambda I - A)^{-1}$ **of order** $m \geq 1$ if λ_0 is an isolated point of the spectrum and
$$B_{-m} \neq 0, \quad B_{-k} = 0, \quad \forall k > m.$$
The following result is proved in Yosida [**113**, Theorem 3, p.299].

THEOREM 3.9. *Let* $A : D(A) \subset X \to X$ *be a closed linear operator in the complex Banach space* X *and let* λ_0 *be a pole of* $(\lambda I - A)^{-1}$ *of order* $m \geq 1$. *Then* λ_0 *is an eigenvalue of* A, *and*
$$R(B_{-1}) = N((\lambda_0 I - A)^n), \ R(I - B_{-1}) = R((\lambda_0 I - A)^n), \ \forall n \geq m,$$
$$X = N((\lambda_0 I - A)^n) \oplus R((\lambda_0 I - A)^n), \ \forall n \geq m.$$

We already knew that $A\mid_{B_{-1}(X)}$ is bounded. Moreover, if λ_0 is a pole of $(\lambda I - A)^{-1}$ of order $m \geq 1$, we have from the above theorem that
$$\left(\lambda_0 I - A\mid_{B_{-1}(X)}\right)^m = 0.$$
From (3.11) for $p = m$, we obtain
$$AB_{-p} = \lambda_0 B_{-p}.$$
Since $B_{-p} \neq 0$, we have $\{\lambda_0\} \subset \sigma\left(A\mid_{B_{-1}(X)}\right)$. To prove the converse inclusion we use the same argument as in the proof of Kato [**63**, Theorem 6.17, p.178]. Set that for $\lambda \in \mathbb{C}$ and let $\varepsilon < |\lambda - \lambda_0|$,
$$L_\lambda = \frac{1}{2\pi i} \int_{S_{\mathbb{C}}(\lambda_0,\varepsilon)^+} \frac{(\lambda' I - A)^{-1}}{\lambda - \lambda'} d\lambda'.$$
Then we have
$$\begin{aligned}(\lambda I - A) L_\lambda &= \frac{1}{2\pi i} \int_{S_{\mathbb{C}}(\lambda_0,\varepsilon)^+} (\lambda I - A) \frac{(\lambda' I - A)^{-1}}{\lambda - \lambda'} d\lambda' \\ &= \frac{1}{2\pi i} \left[\int_{S_{\mathbb{C}}(\lambda_0,\varepsilon)^+} (\lambda' I - A)^{-1} d\lambda' + \int_{S_{\mathbb{C}}(\lambda_0,\varepsilon)^+} \frac{1}{\lambda - \lambda'} d\lambda' \right] \\ &= \frac{1}{2\pi i} \left[\int_{S_{\mathbb{C}}(\lambda_0,\varepsilon)^+} (\lambda' I - A)^{-1} d\lambda' \right] = B_{-1}.\end{aligned}$$
Similarly, we have
$$L_\lambda (\lambda I - A) x = B_{-1} x, \forall x \in D(A).$$
It follows that for each $\lambda \in \mathbb{C} \setminus \{\lambda_0\}$, $\left(\lambda I - A\mid_{B_{-1}(X)}\right)$ is invertible and
$$\left(\lambda I - A\mid_{B_{-1}(X)}\right)^{-1} = L_\lambda \mid_{B_{-1}(X)}.$$
It follows that
$$\sigma\left(A\mid_{B_{-1}(X)}\right) = \{\lambda_0\}.$$
Furthermore, since $\lambda_0 \notin \sigma\left(A\mid_{(I-B_{-1})(X)}\right)$, we have that
$$\sigma\left(A\mid_{(I-B_{-1})(X)}\right) = \sigma(A) \setminus \{\lambda_0\}.$$

Assume that λ_1 and λ_2 are two distinct poles of $(\lambda I - A)^{-1}$. Set for each $i = 1, 2$ that
$$P_i = \frac{1}{2\pi i} \int_{S_{\mathbb{C}}(\lambda_i, \varepsilon)^+} (\lambda I - A)^{-1} d\lambda,$$
where $\varepsilon > 0$ is small enough. It is clear that P_1 commutes with P_2 and
$$P_1 P_2 = P_2 P_1 = 0.$$
Indeed, let $x \in R(P_1)$ be fixed. Since P_1 commutes with $(\lambda I - A)^{-1}$ for each $\lambda \in \rho(A)$, we have
$$P_2 x = \frac{1}{2\pi i} \int_{S_{\mathbb{C}}(\lambda_2, \varepsilon)^+} (\lambda I - A)^{-1} x d\lambda = \frac{1}{2\pi i} \int_{S_{\mathbb{C}}(\lambda_2, \varepsilon)^+} \left(\lambda I - A \mid_{P_1(X)}\right)^{-1} x d\lambda.$$
Furthermore, since $\sigma\left(A \mid_{P_1(X)}\right) = \{\lambda_1\}$, it follows from (3.7) that
$$\begin{aligned} P_2 x &= \frac{1}{2\pi i} \int_{S_{\mathbb{C}}(\lambda_2, \varepsilon)^+} \sum_{n=0}^{\infty} (\lambda - \lambda_2)^n \left[\left(\lambda_2 I - A \mid_{P_1(X)}\right)^{-1}\right]^{n+1} x d\lambda \\ &= \frac{1}{2\pi i} \sum_{n=0}^{\infty} \int_{S_{\mathbb{C}}(\lambda_2, \varepsilon)^+} (\lambda - \lambda_2)^n d\lambda \left[\left(\lambda_2 I - A \mid_{P_1(X)}\right)^{-1}\right]^{n+1} x \\ &= 0. \end{aligned}$$
Hence,
$$P_2 x = 0, \quad \forall x \in R(P_1).$$

ASSUMPTION 3.10. *Let $(X, \|.\|)$ be a complex Banach space and let $A : D(A) \subset X \to X$ be a linear operator satisfying Assumption 2.3. Assume that there exists $\eta \in \mathbb{R}$ such that*
$$\Sigma_\eta := \sigma(A_0) \cap \{\lambda \in \mathbb{C} : \operatorname{Re}(\lambda) > \eta\}$$
is non-empty, finite, and contains only poles of $(\lambda I - A_0)^{-1}$.

By using Lemma 2.1 we know that
$$\sigma(A_0) = \sigma(A),$$
so
$$\Sigma_\eta := \sigma(A) \cap \{\lambda \in \mathbb{C} : Re(\lambda) > \eta\},$$
and for each $\lambda_0 \in \Sigma_\eta$, we set
$$B^0_{\lambda_0, k} = \frac{1}{2\pi i} \int_{S_{\mathbb{C}}(\lambda_0, \varepsilon)^+} (\lambda - \lambda_0)^{-k-1} (\lambda I - A_0)^{-1} d\lambda, \forall k \in \mathbb{Z},$$
and
$$B_{\lambda_0, k} = \frac{1}{2\pi i} \int_{S_{\mathbb{C}}(\lambda_0, \varepsilon)^+} (\lambda - \lambda_0)^{-k-1} (\lambda I - A)^{-1} d\lambda, \forall k \in \mathbb{Z}.$$
We first have the following lemma.

LEMMA 3.11. *Let Assumption 3.10 be satisfied. If $\lambda_0 \in \Sigma_\eta$ is a pole of $(\lambda I - A_0)^{-1}$ of order m, then λ_0 is a pole of order m of $(\lambda I - A)^{-1}$ and*
$$B_{\lambda_0, 1} x = \lim_{\mu \to +\infty} B^0_{\lambda_0, 1} \mu (\mu I - A)^{-1} x, \forall x \in X.$$

PROOF. Let $x \in X$ and $k \in \mathbb{Z}$ be fixed. We have $B_{\lambda_0,k}x \in X_0$, so
$$B_{\lambda_0,k}x = \lim_{\mu \to +\infty} \mu(\mu I - A)^{-1} B_{\lambda_0,k}x.$$
Thus,
$$\begin{aligned}\mu(\mu I - A)^{-1} B_{\lambda_0,k}x &= \frac{1}{2\pi i} \mu(\mu I - A)^{-1} \int_{S_{\mathbb{C}}(\lambda_0,\varepsilon)^+} (\lambda - \lambda_0)^{-k-1} (\lambda I - A)^{-1} x \, d\lambda \\ &= \frac{1}{2\pi i} \int_{S_{\mathbb{C}}(\lambda_0,\varepsilon)^+} (\lambda - \lambda_0)^{-k-1} (\lambda I - A_0)^{-1} \mu(\mu I - A)^{-1} x \, d\lambda \\ &= \lim_{\mu \to +\infty} B^0_{\lambda_0,k} \mu(\mu I - A)^{-1} x,\end{aligned}$$
and the result follows. $\square$

From the above results we immediately have the following result.

THEOREM 3.12. *Let Assumption 3.10 be satisfied. Set*
$$\Pi_0 = \sum_{\lambda_0 \in \Sigma_\eta} B^0_{\lambda_0,-1}, \quad \Pi = \sum_{\lambda_0 \in \Sigma_\eta} B_{\lambda_0,-1}.$$
Then
$$\Pi x = \lim_{\mu \to +\infty} \Pi_0 \mu(\mu I - A)^{-1} x, \forall x \in X.$$
Moreover, we have the following properties:

(i) $\Pi|_{X_0} = \Pi_0$, $\Pi(X) \subset D(A) \subset X_0$, *and*
$$\Pi(\lambda I - A)^{-1} = (\lambda I - A)^{-1} \Pi, \forall \lambda \in \rho(A).$$

(ii) $A|_{\Pi(X)}$ *is bounded*,
$$\sigma\left(A|_{\Pi(X)}\right) = \sigma\left(A_0|_{\Pi_0(X_0)}\right) = \Sigma_\eta,$$
and
$$\sigma\left(A|_{(I-\Pi)(X)}\right) = \sigma\left(A_0|_{(I-\Pi_0)(X_0)}\right) = \sigma(A) \setminus \Sigma_\eta.$$

Let $\widehat{A} : D(\widehat{A}) \subset \widehat{X} \to \widehat{X}$ be the generator of $\{T_{\widehat{A}}(t)\}$, a strongly continuous semigroup of bounded linear operator on a Banach space $\left(\widehat{X}, \|.\|_{\widehat{X}}\right)$. We denote by $\omega_0\left(\widehat{A}\right) \in [-\infty, +\infty)$ the **growth bound of** $\widehat{A}$, which is defined by
$$\omega_0\left(\widehat{A}\right) := \lim_{t \to +\infty} \frac{\ln\left(\|T_{\widehat{A}}(t)\|_{\mathcal{L}(\widehat{X})}\right)}{t},$$
and denote by $\omega_{0,ess}\left(\widehat{A}\right) \in [-\infty, +\infty)$ the **essential growth bound of** $\widehat{A}$, which is defined by
$$\omega_{0,ess}\left(\widehat{A}\right) := \lim_{t \to +\infty} \frac{\ln\left(\tau\left(T_{\widehat{A}}(t) B_{\widehat{X}}(0,1)\right)\right)}{t}$$
where $B_{\widehat{X}}(0,1) = \left\{x \in \widehat{X} : \|x\|_{\widehat{X}} \leq 1\right\}$, and for each bounded set $B \subset \widehat{X}$,
$$\tau(B) = \inf\{\varepsilon > 0 : B \text{ can be covered by a finite number of balls of radius } \leq \varepsilon\}$$
is the Kuratovsky measure of non-compactness.

REMARK 3.13. Note that the existence of the limit in the definition of the growth bound $\omega_0(\widehat{A})$ is proved in Dunford and Schwartz [**40**, Corollary 5, p.619]. The existence of the limit in the definition of the essential growth bound $\omega_{0,ess}(\widehat{A})$ follows from Dunford and Schwartz [**40**, Lemma 4, p.618] and the proof of Webb [**108**, Proposition 4.12, p.170].

The following result is taken from Webb [**108**, Proposition 4.13, p.170-171].

PROPOSITION 3.14. *Let* $\widehat{A} : D(\widehat{A}) \subset \widehat{X} \to \widehat{X}$ *be the generator of* $\{T_{\widehat{A}}(t)\}$, *a strongly continuous semigroup of bounded linear operators on a Banach space* $\left(\widehat{X}, \|.\|_{\widehat{X}}\right)$. *Then*

$$\omega_0\left(\widehat{A}\right) \geq \sup_{\lambda \in \sigma(\widehat{A})} \operatorname{Re}(\lambda), \quad \omega_{0,ess}\left(\widehat{A}\right) \geq \sup_{\lambda \in \sigma_E(\widehat{A})} \operatorname{Re}(\lambda),$$

and

$$\omega_0\left(\widehat{A}\right) = \max\left(\omega_{0,ess}\left(\widehat{A}\right), \sup_{\lambda \in \sigma(\widehat{A}) \setminus \sigma_E(\widehat{A})} \operatorname{Re}(\lambda)\right),$$

where $\sigma_E(\widehat{A})$ *is the essential spectrum of* $\widehat{A}$.

By applying the above result and Proposition 4.11 on p. 166 in Webb [**108**] and Corollary 2.11 on p. 258 in Engel and Nagel [**41**], we obtain the following theorem.

THEOREM 3.15. *Let* $(X, \|.\|)$ *be a complex Banach space and let* $A : D(A) \subset X \to X$ *be a linear operator satisfying Assumption 2.3, and assume that* $\omega_0(A_0) > \omega_{0,ess}(A_0)$. *Then for each* $\eta > \omega_{0,ess}(A_0)$ *such that*

$$\Sigma_\eta := \sigma(A_0) \cap \{\lambda \in \mathbb{C} : \operatorname{Re}(\lambda) \geq \eta\}$$

is nonempty and finite, each $\lambda_0 \in \Sigma_\eta$ *is a pole of* $(\lambda - A_0)^{-1}$ *and* $B^0_{\lambda_0,-1}$ *has a finite rank. Moreover, if we set*

$$\Pi_0 = \sum_{\lambda_0 \in \Sigma_\eta} B^0_{\lambda_0,-1},$$

then

$$\Pi_0 (\lambda - A_0)^{-1} = (\lambda - A_0)^{-1} \Pi_0, \forall \lambda \in \rho(A),$$
$$\omega_0(A_0) = \omega_0\left(A_0 |_{\Pi_0(X)}\right) = \sup_{\lambda \in \Sigma_\eta} \operatorname{Re}(\lambda),$$

and

$$\omega_0\left(A_0 |_{(I-\Pi_0)(X)}\right) \leq \eta.$$

REMARK 3.16. In order to apply the above theorem, we need to check that $\omega_0(A_0) > \omega_{0,ess}(A_0)$. This property can be verified by using perturbation techniques and by applying the results of Thieme [**101**] in the Hille-Yosida case, or the results in Ducrot, Liu and Magal [**38**] in the present context.

CHAPTER 4

Center Manifold Theory

In this chapter, we investigate the existence and smoothness of the center manifold for a nonlinear semiflow $\{U(t)\}_{t\geq 0}$ on X_0, generated by integrated solutions of the Cauchy problem

$$(4.1) \qquad \frac{du(t)}{dt} = Au(t) + F(u(t)), \text{ for } t \geq 0, \text{ with } u(0) = x \in X_0,$$

where $A : D(A) \subset X \to X$ is a linear operator satisfying Assumptions 2.3 and 2.9, and $F : X_0 \to X$ is Lipschitz continuous. So $t \to U(t)x$ is a solution of

$$(4.2) \qquad U(t)x = x + A\int_0^t U(s)x\,ds + \int_0^t F(U(s)x)\,ds, \forall t \geq 0,$$

or equivalently

$$(4.3) \qquad U(t)x = T_{A_0}(t)x + (S_A \diamond F(U(.)x))(t), \forall t \geq 0.$$

This type of problems has been investigated by Thieme [**98**] when A is a Hille-Yosida operator and by Magal and Ruan [**78**] when A satisfies Assumptions 2.3 and 2.9. We know that for each $x \in X_0$, (4.2) has a unique integrated solution $t \to U(t)x$ from $[0, +\infty)$ into X_0. Moreover, the family $\{U(t)\}_{t\geq 0}$ defines a continuous semiflow, that is,

(i) $U(0) = I$ and $U(t)U(s) = U(t+s), \forall t, s \geq 0$,
(ii) The map $(t, x) \to U(t)x$ is continuous from $[0, +\infty) \times X_0$ into X_0.

Furthermore (see Magal and Ruan [**78**]), there exists $\gamma > 0$ such that

$$\|U(t)x - U(t)y\| \leq Me^{\gamma t}\|x - y\|, \quad \forall t \geq 0, \forall x, y \in X_0.$$

Assume that $\overline{x} \in X_0$ is an equilibrium of $\{U(t)\}_{t\geq 0}$ (i.e. $U(t)\overline{x} = \overline{x}, \forall t \geq 0$, or equivalently $\overline{x} \in D(A)$ and $A\overline{x} + F(\overline{x}) = 0$). Then by using (4.2) and by replacing $U(t)x$ by $V(t)x = U(t)x - \overline{x}$, and $F(x)$ by $F(x + \overline{x}) - F(\overline{x})$, without loss of generality we can assume that $\overline{x} = 0$. Moreover, assume that F is differentiable at 0 and denote by $DF(0)$ its differential at 0. Then by using Proposition 2.12 and by replacing A by $A + DF(0)$, and F by $F - DF(0)$, without loss of generality we can also assume that $DF(0) = 0$. So in the sequel, we will assume that we can decompose the space X_0 into X_{0s}, X_{0c}, and X_{0u}, the stable, center, and unstable linear manifold, respectively, corresponding to the spectral decomposition of A_0.

ASSUMPTION 4.1. Assume that Assumption 2.3 and 2.9 are satisfied and there exist two bounded linear projectors with finite rank, $\Pi_{0c} \in \mathcal{L}(X_0) \setminus \{0\}$ and $\Pi_{0u} \in \mathcal{L}(X_0)$, such that

$$\Pi_{0c}\Pi_{0u} = \Pi_{0u}\Pi_{0c} = 0$$

and

$$\Pi_{0k}T_{A_0}(t) = T_{A_0}(t)\Pi_{0k}, \quad \forall t \geq 0, \forall k = \{c, u\}.$$

Assume in addition that

(a) If $\Pi_{0u} \neq 0$, then $\omega_0\left(-A_0\left|_{\Pi_{0u}(X_0)}\right.\right) < 0$.
(b) $\sigma\left(A_0\left|_{\Pi_{0c}(X_0)}\right.\right) \subset i\mathbb{R}$
(c) If $\Pi_{0s} := I - (\Pi_{0c} + \Pi_{0u}) \neq 0$, then $\omega_0\left(A_0\left|_{\Pi_{0s}(X_0)}\right.\right) < 0$.

REMARK 4.2. By Theorem 3.15, Assumption 4.1 is satisfied if and only if

(a) $\omega_{0,ess}(A_0) < 0$.
(b) $\sigma(A_0) \cap i\mathbb{R} \neq \emptyset$.

For each $k = \{c, u\}$, we denote by $\Pi_k : X \to X$ the unique extension of Π_{0k} satisfying (i)-(iii) in Proposition 3.5. Denote

$$\Pi_s = I - (\Pi_c + \Pi_u) \text{ and } \Pi_h = I - \Pi_c.$$

Then we have for each $k \in \{c, h, s, u\}$ that

$$\Pi_k (\lambda I - A)^{-1} = (\lambda I - A)^{-1} \Pi_k, \forall \lambda > \omega,$$
$$\Pi_k (X_0) \subset X_0,$$

and for each $k \in \{c, u\}$ that

$$\Pi_k (X) \subset X_0.$$

For each $k \in \{c, h, s, u\}$, set

$$X_{0k} = \Pi_k(X_0), \ X_k = \Pi_k(X), \ A_k = A\left|_{X_k}\right., \text{ and } A_{0k} = A_0\left|_{X_{0k}}\right..$$

Then for each $k \in \{c, u\}$,

$$X_k = X_{0k}.$$

Thus, by using Lemma 3.6(i) and (3.1) we have for each $k \in \{c, h, s, u\}$ that

$$(A_k)_{\overline{D(A_k)}} = A_0\left|_{X_{0k}}\right. \text{ and } X_{0k} = X_k \cap X_0.$$

In other words, A_{0k} is the part of A_k in $X_{0k} = \overline{D(A_k)}$. Moreover, we have

$$X = X_s \oplus X_c \oplus X_u \text{ and } X_h = X_s \oplus X_u.$$

LEMMA 4.3. *Fix* $\beta \in (0, \min(-\omega_0(A_{0s}), -\omega_0(-A_{0u})))$. *Then we have*

(4.4) $$\|T_{A_{0s}}(t)\|_{\mathcal{L}(X_{0s})} \leq M_s e^{-\beta t}, \forall t \geq 0,$$
(4.5) $$\|e^{-A_{0u}t}\|_{\mathcal{L}(X_{0u})} \leq M_u e^{-\beta t}, \forall t \geq 0$$

with

$$M_s = \sup_{t \geq 0} \|T_{A_{0s}}(t)\|_{\mathcal{L}(X_{0s})} e^{\beta t} < +\infty,$$
$$M_u = \sup_{t \geq 0} \|e^{-A_{0u}t}\|_{\mathcal{L}(X_{0u})} e^{\beta t} < +\infty.$$

Moreover, for each $\eta \in (0, \beta)$, *we have*

(4.6) $$\|e^{A_{0c}t}\|_{\mathcal{L}(X_{0c})} \leq e^{\eta|t|} M_{c,\eta}, \ \forall t \in \mathbb{R},$$

with

$$M_{c,\eta} = \sup_{t \in \mathbb{R}} \|e^{A_{0c}t}\|_{\mathcal{L}(X_{0c})} e^{-\eta|t|} < +\infty.$$

Let $(Y, \|.\|_Y)$ be a Banach space. Let $\eta \in \mathbb{R}$ be a constant and $I \subset \mathbb{R}$ be an interval. Define

$$BC^\eta(I, Y) = \left\{ f \in C(I, Y) : \sup_{t \in I} e^{-\eta|t|} \|f(t)\|_Y < +\infty \right\}.$$

It is well known that $BC^\eta(I, Y)$ is a Banach space when it is endowed with the norm

$$\|f\|_{BC^\eta(I,Y)} = \sup_{t \in I} e^{-\eta|t|} \|f(t)\|_Y.$$

Moreover, the family $\left\{ \left(BC^\eta(I, Y), \|.\|_{BC^\eta(I,Y)} \right) \right\}_{\eta > 0}$ forms a **scale of Banach spaces**, that is, if $0 < \zeta < \eta$ then $BC^\zeta(I, Y) \subset BC^\eta(I, Y)$ and the embedding is continuous; more precisely, we have

$$\|f\|_{BC^\eta(I,Y)} \leq \|f\|_{BC^\zeta(I,Y)}, \ \forall f \in BC^\zeta(I, Y).$$

Let $(Z, \|.\|_Z)$ be a Banach spaces. From now on, we denote by $\text{Lip}(Y, Z)$ (resp. $\text{Lip}_B(Y, Z)$) the space of Lipschitz (resp. Lipschitz and bounded) maps from Y into Z. Set

$$\|F\|_{\text{Lip}(Y,Z)} := \sup_{x, y \in Y : x \neq y} \frac{\|F(x) - F(y)\|_Z}{\|x - y\|_Y}.$$

We shall study the existence and smoothness of center manifolds in the following two sections.

4.1. Existence of center manifolds

In this section, we investigate the existence of center manifolds. From now on we fix $\beta \in (0, \min(-\omega_0(A_{0s}), -\omega_0(-A_{0u})))$. Recall that $u \in C(\mathbb{R}, X_0)$ is **a complete orbit** of $\{U(t)\}_{t \geq 0}$ if

(4.7) $$u(t) = U(t-s)u(s), \ \forall t, s \in \mathbb{R} \text{ with } t \geq s,$$

where $\{U(t)\}_{t \geq 0}$ is a continuous semiflow generated by (4.2).

Note that equation (4.7) is also equivalent to

$$u(t) = u(s) + A \int_0^{t-s} u(s+r) dr + \int_0^{t-s} F(u(s+r)) dr$$

for all $t, s \in \mathbb{R}$ with $t \geq s$, or to

(4.8) $$u(t) = T_{A_0}(t-s) u(s) + (S_A \diamond F(u(s+.))) (t-s)$$

for each $t, s \in \mathbb{R}$ with $t \geq s$.

DEFINITION 4.4. Let $\eta \in (0, \beta)$. The η- **center manifold** of (4.1), denoted by V_η, is the set of all points $x \in X_0$, such that there exists $u \in BC^\eta(\mathbb{R}, X_0)$, a complete orbit of $\{U(t)\}_{t \geq 0}$, such that $u(0) = x$.

Let $u \in BC^\eta(\mathbb{R}, X_0)$. For all $\tau \in \mathbb{R}$, we have

$$e^{-\eta|\tau|} \|u\|_{BC^\eta(\mathbb{R},X_0)} \leq \|u(.+\tau)\|_{BC^\eta(\mathbb{R},X_0)} \leq e^{\eta|\tau|} \|u\|_{BC^\eta(\mathbb{R},X_0)}.$$

So for each $\eta > 0$, V_η is invariant under the semiflow $\{U(t)\}_{t \geq 0}$, that is,

$$U(t) V_\eta = V_\eta, \ \forall t \geq 0.$$

Moreover, we say that $\{U(t)\}_{t\geq 0}$ is **reduced on** V_η if there exists a map $\Psi^\eta : X_{0c} \to X_{0h}$ such that
$$V_\eta = \text{Graph}(\Psi) = \{x_c + \Psi(x_c) : x_c \in X_{0c}\}.$$

Before proving the main results of this chapter, we need some preliminary lemmas.

LEMMA 4.5. *Let Assumption 4.1 be satisfied. Let $\tau > 0$ be fixed. Then for each $f \in C([0,\tau], X)$ and each $t \in [0,\tau]$, we have*

(4.9) $\qquad \Pi_{0s}(S_A \diamond f)(t) = (S_A \diamond \Pi_s f)(t) = (S_{A_s} \diamond \Pi_s f)(t),$

and for each $k \in \{c, u\}$,

(4.10) $\quad \Pi_{0k}(S_A \diamond f)(t) = (S_A \diamond \Pi_k f)(t) = \int_0^t e^{A_{0k}(t-r)} \Pi_k f(r) dr, \ \forall t \in [0,\tau].$

Furthermore, for each $\gamma > -\beta$, there exists $\widehat{C}_{s,\gamma} > 0$, such that for each $f \in C([0,\tau], X)$ and each $t \in [0,\tau]$, we have

(4.11) $\qquad e^{-\gamma t} \|\Pi_{0s}(S_A \diamond f)(t)\| \leq \widehat{C}_{s,\gamma} \sup_{s \in [0,t]} e^{-\gamma s} \|f(s)\| ds.$

PROOF. The first part (i.e. equations (4.9) and (4.10)) of the lemma is a consequence of Proposition 3.7. Moreover, applying Proposition 2.13 to $(S_{A_s} \diamond \Pi_s f)(t)$ and using (4.4), we obtain (4.11). $\square$

LEMMA 4.6. *Let Assumption 4.1 be satisfied. Then we have the following:*

(i) *For each $\eta \in [0, \beta)$, each $f \in BC^\eta(\mathbb{R}, X)$, and each $t \in \mathbb{R}$,*
$$K_s(f)(t) := \lim_{r \to -\infty} \Pi_{0s}(S_A \diamond f(r + .))(t - r) \text{ exists.}$$

(ii) *For each $\eta \in [0, \beta)$, K_s is a bounded linear operator from $BC^\eta(\mathbb{R}, X)$ into $BC^\eta(\mathbb{R}, X_{0s})$. More precisely, for each $\nu \in (-\beta, 0)$, we have*
$$\|K_s\|_{\mathcal{L}(BC^\eta(\mathbb{R},X), BC^\eta(\mathbb{R}, X_{0s}))} \leq \widehat{C}_{s,\nu}, \forall \eta \in [0, -\nu],$$
where $\widehat{C}_{s,\nu} > 0$ is the constant introduced in (4.11).

(iii) *For each $\eta \in [0, \beta)$, each $f \in BC^\eta(\mathbb{R}, X)$, and each $t, s \in \mathbb{R}$ with $t \geq s$,*
$$K_s(f)(t) - T_{A_{0s}}(t-s) K_s(f)(s) = \Pi_{0s}(S_A \diamond f(s + .))(t - s).$$

PROOF. (i) and (iii) Let $\eta \in (0, \beta)$ be fixed. By using (2.7), we have for each $t, s, r \in \mathbb{R}$ with $r \leq s \leq t$, and each $f \in BC^\eta(\mathbb{R}, X)$ that
$$(S_A \diamond f(r + .))(t - r) = T_{A_0}(t - s)(S_A \diamond f(r + .))(s - r) + (S_A \diamond f(s + .))(t - s).$$

By projecting this equation on X_{0s}, we obtain for all $t, s, r \in \mathbb{R}$ with $r \leq s \leq t$ that

(4.12)
$$\begin{aligned}\Pi_{0s}(S_A \diamond f(r + .))(t - r) \\ = T_{A_{0s}}(t - s)\Pi_{0s}(S_A \diamond f(r + .))(s - r) \\ + \Pi_{0s}(S_A \diamond f(s + .))(t - s).\end{aligned}$$

4.1. EXISTENCE OF CENTER MANIFOLDS

Let $\nu \in (-\beta, -\eta)$ be fixed. Then by using (4.4) and (4.11), we have for all $t, s, r \in \mathbb{R}$ with $r \leq s \leq t$ that

$$\|\Pi_{0s}\left(S_A \diamond f(r + .)\right)(t - r) - \Pi_{0s}\left(S_A \diamond f(s + .)\right)(t - s)\|$$
$$= \|T_{A_{0s}}(t - s)\Pi_{0s}\left(S_A \diamond f(r + .)\right)(s - r)\|$$
$$\leq M_s e^{-\beta(t-s)} \widehat{C}_{s,\nu} e^{\nu(s-r)} \sup_{l \in [0, s-r]} e^{-\nu l} \|f(r + l)\|$$
$$= M_s \widehat{C}_{s,\nu} e^{-\beta(t-s)} e^{\nu(s-r)} \sup_{\sigma \in [r,s]} e^{-\nu(\sigma - r)} \|f(\sigma)\|$$
$$= M_s \widehat{C}_{s,\nu} e^{-\beta(t-s)} e^{\nu s} \sup_{l \in [r,s]} e^{-\nu \sigma} e^{\eta |\sigma|} e^{-\eta |\sigma|} \|f(\sigma)\|$$
$$\leq \|f\|_{BC^\eta(\mathbb{R}, X)} M_s \widehat{C}_{s,\nu} e^{-\beta(t-s)} e^{\nu s} \sup_{\sigma \in [r,s]} e^{-\nu \sigma} e^{\eta |\sigma|}.$$

Hence, for all $s, r \in \mathbb{R}_-$ with $s \geq r$, we obtain

$$\|\Pi_{0s}\left(S_A \diamond f(r + .)\right)(t - r) - \Pi_{0s}\left(S_A \diamond f(s + .)\right)(t - s)\|$$
$$\leq \|f\|_{BC^\eta(\mathbb{R}, X)} M_s \widehat{C}_{s,\nu} e^{-\beta(t-s)} e^{\nu s} \sup_{\sigma \in [r,s]} e^{-(\nu+\eta)\sigma}.$$

Because $-(\nu + \eta) > 0$, we have

$$\|\Pi_{0s}\left(S_A \diamond f(r + .)\right)(t - r) - \Pi_{0s}\left(S_A \diamond f(s + .)\right)(t - s)\|$$
$$\leq \|f\|_{BC^\eta(\mathbb{R}, X)} M_s \widehat{C}_{s,\nu} e^{-\beta(t-s)} e^{\nu s} e^{-(\nu+\eta)s}$$
$$= \|f\|_{BC^\eta(\mathbb{R}, X)} M_s \widehat{C}_{s,\nu} e^{-\beta t} e^{(\beta - \eta)s}.$$

Since $\beta - \eta > 0$, by using Cauchy sequences, we deduce that

$$K_s(f)(t) = \lim_{s \to -\infty} \Pi_{0s}\left(S_A \diamond f(s + .)\right)(t - s) \text{ exists}.$$

Taking the limit as r goes to $-\infty$ in (4.12), we obtain (iii).

(ii) Let $\nu \in (-\beta, 0)$ and $\eta \in [0, -\nu]$ be fixed. For each $f \in BC^\eta(\mathbb{R}, X)$ and each $t \in \mathbb{R}$, we have

$$\begin{aligned}
\|K_s(f)(t)\| &= \lim_{r \to -\infty} \|\Pi_{0s}\left(S_A \diamond f(r + .)\right)(t - r)\| \\
&\leq \widehat{C}_{s,\nu} \limsup_{r \to -\infty} e^{\nu(t-r)} \sup_{l \in [0, t-r]} e^{-\nu l} \|f(r + l)\| \\
&= \widehat{C}_{s,\nu} \limsup_{r \to -\infty} e^{\nu(t-r)} \sup_{\sigma \in [r,t]} e^{-\nu(\sigma - r)} \|f(\sigma)\| \\
&= \widehat{C}_{s,\nu} \limsup_{r \to -\infty} e^{\nu t} \sup_{\sigma \in [r,t]} e^{-\nu \sigma} e^{\eta |\sigma|} e^{-\eta |\sigma|} \|f(\sigma)\| \\
&= \widehat{C}_{s,\nu} e^{\nu t} \|f\|_\eta \sup_{\sigma \in (-\infty, t]} e^{-\nu \sigma} e^{\eta |\sigma|}.
\end{aligned}$$

Since $(\nu + \eta) \leq 0$, we deduce that if $t \leq 0$,

$$\begin{aligned}
e^{-\eta |t|} \|K_s(f)(t)\| &\leq \widehat{C}_{s,\nu} e^{(\nu+\eta)t} \|f\|_\eta \sup_{\sigma \in (-\infty, t]} e^{-(\nu+\eta)\sigma} = \widehat{C}_{s,\nu} e^{(\nu+\eta)t} \|f\|_\eta e^{-(\nu+\eta)t} \\
&= \widehat{C}_{s,\nu} \|f\|_\eta
\end{aligned}$$

and since $(\eta - \nu) > 0$, it follows that if $t \geq 0$,

$$\begin{aligned}
e^{-\eta|t|} \|K_s(f)(t)\| &\leq \widehat{C}_{s,\nu} e^{(\nu-\eta)t} \|f\|_\eta \sup_{\sigma \in (-\infty,t]} e^{-\nu\sigma} e^{\eta|\sigma|} \\
&\leq \widehat{C}_{s,\nu} \|f\|_\eta e^{(\nu-\eta)t} \max(\sup_{\sigma \in (-\infty,0]} e^{-(\nu+\eta)\sigma}, \sup_{\sigma \in [0,t]} e^{(\eta-\nu)\sigma}) \\
&= \widehat{C}_{s,\nu} \|f\|_\eta e^{(\nu-\eta)t} e^{(\eta-\nu)t} = \widehat{C}_{s,\nu} \|f\|_\eta.
\end{aligned}$$

This completes the proof. $\square$

LEMMA 4.7. *Let Assumption 4.1 be satisfied. Let $\eta \in [0, \beta)$ be fixed. Then we have the following:*

(i) *For each $f \in BC^\eta(\mathbb{R}, X)$ and each $t \in \mathbb{R}$,*

$$K_u(f)(t) := -\int_t^{+\infty} e^{-A_{0u}(l-t)} \Pi_u f(l) dl := -\lim_{r \to +\infty} \int_t^r e^{-A_{0u}(l-t)} \Pi_u f(l) dl$$

exists.

(ii) *K_u is a bounded linear operator from $BC^\eta(\mathbb{R}, X)$ into $BC^\eta(\mathbb{R}, X_{0u})$ and*

$$\|K_u\|_{\mathcal{L}(BC^\eta(\mathbb{R}, X))} \leq \frac{M_u \|\Pi_u\|_{\mathcal{L}(X)}}{(\beta - \eta)}.$$

(iii) *For each $f \in BC^\eta(\mathbb{R}, X)$ and each $t, s \in \mathbb{R}$ with $t \geq s$,*

$$K_u(f)(t) - e^{A_{0u}(t-s)} K_u(f)(s) = \Pi_{0u}(S_A \diamond f(s + \cdot))(t-s).$$

PROOF. By using (4.5) and the same argument as in the proof of Lemma 4.6, we obtain (i) and (ii). Moreover, for each $s, t, r \in \mathbb{R}$ with $s \leq t \leq r$, we have

$$\begin{aligned}
\int_s^r e^{A_{0u}(s-l)} \Pi_u f(l) dl &= \int_s^t e^{A_{0u}(s-l)} \Pi_u f(l) dl + \int_t^r e^{A_{0u}(s-l)} \Pi_u f(l) dl \\
&= \int_s^t e^{A_{0u}(s-l)} \Pi_u f(l) dl + e^{A_{0u}(s-t)} \int_t^r e^{A_{0u}(t-l)} \Pi_u f(l) dl.
\end{aligned}$$

It follows that

$$e^{A_{0u}(t-s)} \int_s^r e^{A_{0u}(s-l)} \Pi_u f(l) dl = \int_s^t e^{A_{0u}(t-l)} \Pi_u f(l) dl + \int_t^r e^{A_{0u}(t-l)} \Pi_u f(l) dl.$$

When $r \to +\infty$, we obtain for all $s, t \in \mathbb{R}$ with $s \leq t$ that

$$\begin{aligned}
-e^{A_{0u}(t-s)} K_{u,\eta}(f)(s) &= \int_0^{t-s} e^{A_{0u}(t-s-r)} \Pi_u f(s+r) dr - K_{u,\eta}(f)(t) \\
&= \Pi_u(S_A \diamond f(s + \cdot))(t-s) - K_{u,\eta}(f)(t).
\end{aligned}$$

This gives (iii). $\square$

LEMMA 4.8. *Let Assumption 4.1 be satisfied. Let $\eta \in (0, \beta)$ be fixed. For each $x_c \in X_{0c}$, each $f \in BC^\eta(\mathbb{R}, X)$, and each $t \in \mathbb{R}$, denote*

$$K_1(x_c)(t) := e^{A_{0c}t} x_c, \quad K_c(f)(t) := \int_0^t e^{A_{0c}(t-s)} \Pi_c f(s) ds.$$

Then K_1 (respectively K_c) is a bounded linear operator from X_{0c} into $BC^\eta(\mathbb{R}, X_{0c})$ (respectively from $BC^\eta(\mathbb{R}, X)$ into $BC^\eta(\mathbb{R}, X_{0c})$), and

$$\|K_1\|_{\mathcal{L}(X_c, BC^\eta(\mathbb{R}, X))} \leq \max\left(\sup_{t \geq 0}\left\|e^{(A_c - \eta I)t}\right\|, \sup_{t \geq 0}\left\|e^{-(A_c + \eta I)t}\right\|\right),$$

$$\|K_c\|_{\mathcal{L}(BC^\eta(\mathbb{R}, X))} \leq \|\Pi_c\|_{\mathcal{L}(X)} \max\left(\int_0^{+\infty}\left\|e^{(A_c - \eta I)l}\right\|dl, \int_0^{+\infty}\left\|e^{-(A_c + \eta I)l}\right\|dl\right).$$

PROOF. The proof is straightforward. □

LEMMA 4.9. *Let Assumption 4.1 be satisfied. Let $\eta \in (0, \beta)$ and $u \in BC^\eta(\mathbb{R}, X_0)$ be fixed. Then u is a complete orbit of $\{U(t)\}_{t \geq 0}$ if and only if for each $t \in \mathbb{R}$,*

$$(4.13) \quad \begin{aligned} u(t) = &K_1(\Pi_{0c}u(0))(t) + K_c(F(u(.)))(t) \\ &+ K_u(F(u(.)))(t) + K_s(F(u(.)))(t). \end{aligned}$$

PROOF. Let $u \in BC^\eta(\mathbb{R}, X_0)$. Assume first that u is a complete orbit of $\{U(t)\}_{t \geq 0}$. Then for $k \in \{c, u\}$ we have for all $l, r \in \mathbb{R}$ with $r \leq l$ that

$$\Pi_{0k}u(l) = e^{A_{0k}(l-r)}\Pi_{0k}u(r) + \int_r^l e^{A_{0k}(l-s)}\Pi_k F(u(s))ds.$$

Thus,

$$\frac{d\Pi_{0k}u(t)}{dt} = A_{0k}\Pi_{0k}u(t) + \Pi_k F(u(t)), \quad \forall t \in \mathbb{R}.$$

From this ordinary differential equation, we first deduce that

$$(4.14) \quad \Pi_{0c}u(t) = e^{A_{0c}t}\Pi_{0c}u(0) + \int_0^t e^{A_{0c}(t-s)}\Pi_c F(u(s))ds, \forall t \in \mathbb{R}.$$

Hence, for each $t, l \in \mathbb{R}$,

$$\Pi_{0u}u(t) = e^{A_{0u}(t-l)}\Pi_{0u}u(l) + \int_l^t e^{A_{0u}(t-s)}\Pi_u F(u(s))ds, \forall t, l \in \mathbb{R}.$$

It follows that for each $l \geq 0$,

$$\left\|e^{-A_{0u}(l-t)}\Pi_{0u}u(l)\right\| \leq M_u \|\Pi_u\|_{\mathcal{L}(X)} e^{-\beta(l-t)} e^{\eta l} \|u\|_{BC^\eta(\mathbb{R}, X_0)}.$$

So when l goes to $+\infty$, we obtain

$$(4.15) \quad \Pi_{0u}u(t) = -\int_t^{+\infty} e^{A_{0u}(t-s)}\Pi_u F(u(s))ds, \quad \forall t \in \mathbb{R}.$$

Furthermore, we have for all $t, l \in \mathbb{R}$ with $t \geq l$ that

$$\Pi_{0s}u(t) = T_{A_{0s}}(t-l)\Pi_{0s}u(l) + \Pi_{0s}\left(S_A \diamond F(u(l+.))\right)(t-l)$$

and for each $l \leq 0$ that

$$\|T_{A_{0s}}(t-l)\Pi_{0s}u(l)\| \leq e^{-\beta t} M_s \|u\|_\eta e^{(\beta - \eta)l}.$$

Taking $l \to -\infty$, we obtain

$$(4.16) \quad \Pi_{0s}u(t) = K_{s,\eta}\left(F(u(.))\right)(t), \quad \forall t \in \mathbb{R}.$$

Finally, summing up (4.14), (4.15), and (4.16), we obtain (4.13).

Conversely, assume that u is a solution of (4.13). Then

$$\Pi_{0c}u(t) = e^{A_{0c}t}\Pi_{0c}u(0) + \int_0^t e^{A_{0c}(t-s)}\Pi_c F(u(s))ds, \quad \forall t \in \mathbb{R}.$$

It follows that
$$\frac{d\Pi_{0c}u(t)}{dt} = A_{0c}\Pi_{0c}u(t) + \Pi_c F(u(t)), \quad \forall t \in \mathbb{R}.$$
Thus, for $l, r \in \mathbb{R}_-$ with $r \leq l$,
$$\Pi_{0c}u(l) = T_{A_0}(t-s)\Pi_{0c}u(r) + \Pi_{0c}\left(S_A \diamond F(u(s+.))\right)(t-s).$$
Moreover, using Lemma 4.6 (iii) and Lemma 4.7 (iii), we deduce that for all $t, s \in \mathbb{R}$ with $t \geq s$
$$\Pi_{0s}u(t) - T_{A_0}(t-s)\Pi_{0s}u(s) = \Pi_{0s}\left(S_A \diamond F(u(s+.))\right)(t-s),$$
$$\Pi_{0u}u(t) - T_{A_0}(t-s)\Pi_{0u}u(s) = \Pi_{0u}\left(S_A \diamond F(u(s+.))\right)(t-s).$$
Therefore, u satisfies (4.8) and is a complete orbit of $\{U(t)\}_{t \geq 0}$. $\square$

Let $\eta \in (0, \beta)$ be fixed. We rewrite equation (4.13) as the following fixed point problem: To find $u \in BC^\eta(\mathbb{R}, X)$ such that
$$(4.17) \qquad u = K_1(\Pi_{0c}u(0)) + K_2 \Phi_F(u),$$
where the nonlinear operator $\Phi_F \in \mathrm{Lip}\left(BC^\eta(\mathbb{R}, X_0), BC^\eta(\mathbb{R}, X)\right)$ is defined by
$$\Phi_F(u)(t) = F(u(t)), \quad \forall t \in \mathbb{R}$$
and $K_2 \in \mathcal{L}\left(BC^\eta(\mathbb{R}, X), BC^\eta(\mathbb{R}, X_0)\right)$ is the linear operator defined by
$$K_2 = K_c + K_s + K_u.$$
Moreover, we have the following estimates
$$\|K_1\|_{\mathcal{L}(X_c, BC^\eta(\mathbb{R}, X))} \leq \max(\sup_{t \geq 0}\left\|e^{(A_c - \eta Id)t}\right\|, \sup_{t \geq 0}\left\|e^{-(A_c + \eta Id)t}\right\|),$$
$$\|\Phi_F\|_{\mathrm{Lip}} \leq \|F\|_{\mathrm{Lip}},$$
and for each $\nu \in (-\beta, 0)$, we have
$$\|K_2\|_{\mathcal{L}(BC^\eta(\mathbb{R}, X))} \leq \gamma(\nu, \eta), \forall \eta \in (0, -\nu],$$
where
$$(4.18) \quad \begin{aligned}\gamma(\nu, \eta) &:= \widehat{C}_{s,\nu} + \frac{M_u \|\Pi_u\|_{\mathcal{L}(X)}}{(\beta - \eta)} \\ &\quad + \|\Pi_c\|_{\mathcal{L}(X)} \max\left(\int_0^{+\infty}\left\|e^{(A_c - \eta Id)l}\right\| dl, \int_0^{+\infty}\left\|e^{-(A_c + \eta)l}\right\| dl\right).\end{aligned}$$
Moreover, by Lemma 4.9, the η-center manifold is given by
$$(4.19) \quad V_\eta = \{x \in X_0 : \exists u \in BC^\eta(\mathbb{R}, X_0) \text{ a solution of } (4.17) \text{ and } u(0) = x\}.$$

We are now in the position to prove the existence of center manifolds for semilinear equations with non-dense domain, which is a generalization of Vanderbauwhede and Iooss [106, Theorem 1, p.129].

THEOREM 4.10. *Let Assumption 4.1 be satisfied. Let $\eta \in (0, \beta)$ be fixed and let $\delta_0 = \delta_0(\eta) > 0$ be such that*
$$\delta_0 \|K_2\|_{\mathcal{L}(BC^\eta(\mathbb{R}, X))} < 1.$$
Then for each $F \in \mathrm{Lip}(X_0, X)$ with $\|F\|_{\mathrm{Lip}(X_0, X)} \leq \delta_0$, there exists a Lipschitz continuous map $\Psi : X_{0c} \to X_{0h}$ such that
$$V_\eta = \{x_c + \Psi(x_c) : x_c \in X_{0c}\}.$$
Moreover, we have the following properties:

4.1. EXISTENCE OF CENTER MANIFOLDS

(i) $\sup_{x_c \in X_c} \|\Psi(x_c)\| \le \|K_s + K_u\|_{\mathcal{L}(BC^\eta(\mathbb{R},X))} \sup_{x \in X_0} \|\Pi_h F(x)\|$.

(ii)

(4.20) $\|\Psi\|_{\text{Lip}(X_{0c}, X_{0h})} \le \dfrac{\|K_s + K_u\|_{\mathcal{L}(BC^\eta(\mathbb{R},X))} \|F\|_{\text{Lip}(X_0, X)} \|K_1\|_{\mathcal{L}(X_c, BC^\eta(\mathbb{R}, X_0))}}{1 - \|K_2\|_{\mathcal{L}(BC^\eta(\mathbb{R},X))} \|F\|_{\text{Lip}(X_0, X)}}$.

(iii) *For each $u \in C(\mathbb{R}, X_0)$, the following statement are equivalent:*
  (1) *$u \in BC^\eta(\mathbb{R}, X_0)$ is a complete orbit of $\{U(t)\}_{t \ge 0}$.*
  (2) *$\Pi_{0h} u(t) = \Psi(\Pi_{0c} u(t)), \forall t \in \mathbb{R}$, and $\Pi_{0c} u(.) : \mathbb{R} \to X_{0c}$ is a solution of the ordinary differential equation*

(4.21) $$\dfrac{dx_c(t)}{dt} = A_{0c} x_c(t) + \Pi_c F[x_c(t) + \Psi(x_c(t))].$$

PROOF. (i) Since $\|F\|_{\text{Lip}} \|K_2\|_{\mathcal{L}(BC^\eta(\mathbb{R},X))} < 1$, the map $(Id - K_2 \Phi_F)$ is invertible, $(Id - K_2 \Phi_F)^{-1}$ is Lipschitz continuous, and

(4.22) $\left\|(Id - K_2 \Phi_F)^{-1}\right\|_{\text{Lip}(BC^\eta(\mathbb{R}, X_0))} \le \dfrac{1}{1 - \|K_2\|_{\mathcal{L}(BC^\eta(\mathbb{R},X))} \|F\|_{\text{Lip}(X_0, X)}}$.

Let $x \in X_0$ be fixed. By Lemma 4.9, we know that $x \in V_\eta$ if and only if there exists $u_{\Pi_{0c} x} \in BC^\eta(\mathbb{R}, X)$, such that $u_{\Pi_{0c} x}(0) = x$ and

$$u_{\Pi_{0c} x} = K_1(\Pi_{0c} x) + K_2 \Phi_F(u_{\Pi_{0c} x}).$$

So

$$V_\eta = \left\{ (Id - K_2 \Phi_F)^{-1} K_1(x_c)(0) : x_c \in X_{0c} \right\}.$$

We define $\Psi : X_{0c} \to X_{0h}$ by

(4.23) $\Psi(x_c) = \Pi_{0h}(Id - K_2 \Phi_F)^{-1} K_1(x_c)(0), \forall x_c \in X_{0c}$.

Then

$$V_\eta = \{x_c + \Psi(x_c) : x_c \in X_{0c}\}.$$

For each $x_c \in X_{0c}$, set

$$u_{x_c} = (Id - K_2 \Phi_F)^{-1} K_1(x_c).$$

We have

$$u_{x_c} = K_1(x_c) + K_2 \Phi_F(u_{x_c}).$$

By projecting on X_{0h}, we obtain

$$\Pi_{0h} u_{x_c} = [K_s + K_u] \Phi_F(u_{x_c}),$$

so

(4.24) $$\Psi(x_c) = [K_s + K_u] \Phi_F(u_{x_c})(0)$$

and (i) follows.

(ii) It follows from (4.22) and (4.24).

(iii) Assume first that $u \in BC^\eta(\mathbb{R}, X_0)$ is a complete orbit of $\{U(t)\}_{t \ge 0}$. Then by the definition of V_η, we have $u(t) \in V_\eta, \forall t \in \mathbb{R}$. Hence,

$$\Pi_{0h} u(t) = \Psi(\Pi_{0c} u(t)), \quad \forall t \in \mathbb{R}.$$

Moreover, by projecting (4.8) on X_{0c}, we have for each $t, s \in \mathbb{R}$ with $t \ge s$ that

$$\Pi_{0c} u(t) = e^{A_{0c}(t-s)} \Pi_{0c} u(s) + \int_0^{t-s} e^{A_{0c}(t-s-l)} \Pi_c F(u(s+l)) \, dl.$$

Thus, $t \to \Pi_{0c} u(t)$ is a solution of (4.21).

Conversely assume that $u \in C(\mathbb{R}, X_0)$ satisfies (iii)(2). Then
$$\Pi_{0h} u(t) = \Psi(\Pi_{0c} u(t)), \quad \forall t \in \mathbb{R},$$
and $\Pi_{0c} u(.) : \mathbb{R} \to X_{0c}$ is a solution of (4.21). Set $x = u(0)$. We know that $x \in V_\eta$, and by the definition of V_η, there exists $v \in BC^\eta(\mathbb{R}, X_0)$, a complete orbit of $\{U(t)\}_{t \geq 0}$, such that $v(0) = x$. But since V_η is invariant under the semiflow, we deduce that
$$\Pi_{0h} v(t) = \Psi(\Pi_{0c} v(t)), \quad \forall t \in \mathbb{R},$$
and $\Pi_{0c} v(.) : \mathbb{R} \to X_{0c}$ is a solution of (4.21). Finally, since $\Pi_{0c} v(0) = \Pi_{0c} u(0)$, and since F and Ψ are Lipschitz continuous, we deduce that (4.21) has at most one solution. It follows that
$$\Pi_{0c} v(t) = \Pi_{0c} u(t), \forall t \in \mathbb{R},$$
and by construction
$$\Pi_{0h} v(t) = \Psi(\Pi_{0c} v(t)) = \Psi(\Pi_{0c} u(t)) = \Pi_{0h} u(t), \quad \forall t \in \mathbb{R}.$$
Thus,
$$u(t) = v(t), \quad \forall t \in \mathbb{R}.$$
Therefore, $u \in BC^\eta(\mathbb{R}, X_0)$ is a complete orbit of $\{U(t)\}_{t \geq 0}$. $\square$

PROPOSITION 4.11. *Let Assumption 4.1 be satisfied. Assume in addition that $F \in \mathrm{Lip}_B(X_0, X)$ (i.e. F is Lipschitz and bounded). Then*
$$V_\eta = V_\xi, \quad \forall \eta, \xi \in (0, \beta).$$

PROOF. Let $\eta, \xi \in (0, \beta)$ be such that $\xi < \eta$. Let $x \in V_\xi$. By the definition of V_ξ there exists $u \in BC^\xi(\mathbb{R}, X_0)$, a complete orbit of $\{U(t)\}_{t \geq 0}$, such that $u(0) = x$. But $BC^\xi(\mathbb{R}, X_0) \subset BC^\eta(\mathbb{R}, X_0)$, so $u \in BC^\eta(\mathbb{R}, X_0)$, and we deduce that $x \in V_\eta$.

Conversely, let $x \in V_\eta$ be fixed. By the definition of V_η there exists $u \in BC^\eta(\mathbb{R}, X_0)$, a complete orbit of $\{U(t)\}_{t \geq 0}$, such that $u(0) = x$. By Lemma 4.9 we deduce that u is a solution of
$$u = K_1(\Pi_{0c} u(0)) + K_2 \Phi_F(u).$$
But $K_1(\Pi_{0c} u(0)) \in BC^\xi(\mathbb{R}, X_0)$ and F is bounded, so we have $\Phi_F(u) \in BC^0(\mathbb{R}, X_0) \subset BC^\xi(\mathbb{R}, X_0)$ and
$$K_2 \Phi_F(u) \in BC^\xi(\mathbb{R}, X_0).$$
Hence, $u \in BC^\xi(\mathbb{R}, X_0)$ and
$$u = K_1(\Pi_{0c} u(0)) + K_2 \Phi_F(u).$$
Using again Lemma 4.9 once more, we obtain that $x \in V_\xi$. $\square$

4.2. Smoothness of center manifolds

In the sequel, we will use the following notation. Let $k \geq 1$ be an integer, let $Y_1, Y_2, .., Y_k, Y$ and Z be Banach spaces, let V be an open subset of Y. Denote $\mathcal{L}^{(k)}(Y_1, Y_2, .., Y_k, Z)$ (resp. $\mathcal{L}^{(k)}(Y, Z)$) the space of bounded k-linear maps from $Y_1 \times ... \times Y_k$ into Z (resp. from Y^k into Z). Let $W \in C^k(V, Z)$ be fixed. We choose the convention that if $l = 1, ..., k-1$ and $u, \widehat{u} \in V$ with $u \neq \widehat{u}$, the quantity

$$\sup_{u_1, ..., u_l \in B_Y(0,1)} \frac{\left\| \left[D^l W(u) - D^l W(\widehat{u}) \right](u_1, ..., u_l) - D^{l+1} W(\widehat{u})(u - \widehat{u}, u_1, ..., u_l) \right\|}{\|u - \widehat{u}\|}$$

4.2. SMOOTHNESS OF CENTER MANIFOLDS

goes to 0 as $\|u - \widehat{u}\| \to 0$. Set

$$C_b^k(V, Z) := \left\{ W \in C^k(V, Z) : |W|_{j,V} := \sup_{x \in V} \|D^j W(x)\| < +\infty, \ 0 \leq j \leq k \right\}.$$

For each $\eta \in [0, \beta)$, consider $K_h : BC^\eta(\mathbb{R}, X) \to BC^\eta(\mathbb{R}, X_{0h})$, the bounded linear operator defined by

$$K_h = K_s + K_u,$$

where K_s and K_u are the bounded linear operators defined, respectively, in Lemma 4.6 and Lemma 4.7. For each $\varrho > 0$ and each $\eta \geq 0$, set

$$V_\varrho := \{ x \in X_0 : \|\Pi_h x\| < \varrho \}, \quad \overline{V}_\varrho := \{ x \in X_0 : \|\Pi_h x\| \leq \varrho \},$$

and

$$\overline{V}_\varrho^\eta := \left\{ u \in BC^\eta(\mathbb{R}, X_0) : u(t) \in \overline{V}_\varrho, \forall t \in \mathbb{R} \right\}.$$

Note that since $\overline{V}_\varrho$ is a closed and convex subset of X_0, so is $\overline{V}_\varrho^\eta$ for each $\eta \geq 0$.
We make the following assumption.

ASSUMPTION 4.12. *Let $k \geq 1$ be an integer and let $\eta, \widehat{\eta} \in \left(0, \frac{\beta}{k} \right)$ such that $k\eta < \widehat{\eta} < \beta$. Assume*
   a) $F \in \text{Lip}(X_0, X) \cap C_b^k(V_\varrho, X)$;
   b) $\varrho_0 := \|K_h\|_{\mathcal{L}(BC^0(\mathbb{R}, X))} \|\Pi_h F\|_{0, X_0} < \varrho$;
   c) $\sup_{\theta \in [\eta, \widehat{\eta}]} \|K_2\|_{\mathcal{L}(BC^\theta(\mathbb{R}, X))} \|F\|_{\text{Lip}(X_0, X)} < 1$.

Note that by using (4.18) we deduce that

$$\sup_{\theta \in [\eta, \widehat{\eta}]} \|K_2\|_{\mathcal{L}(BC^\theta(\mathbb{R}, X))} < +\infty.$$

Thus, Assumption 4.12 makes sense.

Following the approach of Vanderbauwhede [104, Corollary 3.6] and Vanderbauwhede and Iooss [106, Theorem 2], we obtain the following result on the smoothness of center manifolds.

THEOREM 4.13. *Let Assumptions 4.1 and 4.12 be satisfied. Then the map Ψ given by Theorem 4.10 belongs to the space $C_b^k(X_c, X_h)$.*

The above result was stated without proof in [106, Theorem 2]. For the sake of completeness we now prove Theorem 4.13. We first need some preliminary results.

DEFINITION 4.14. *Let X be a metric space and $H : X \to X$ be a map. A fixed point $\overline{x} \in X$ of H is said to be* **attracting** *if*

$$\lim_{n \to +\infty} H^n(x) = \overline{x} \quad \text{for each } x \in X.$$

The following lemma is an extension of the Fibre contraction theorem (which corresponds to the case $k = 1$). This result is taken from [104, Corollary 3.6].

LEMMA 4.15. *Let $k \geq 1$ be an integer and let $(M_0, d_0), (M_1, d_1), ..., (M_k, d_k)$ be complete metric spaces. Let $H : M_0 \times M_1 \times ... \times M_k \to M_0 \times M_1 \times ... \times M_k$ be a mapping of the form*

$$H(x_0, x_1, ..., x_k) = (H_0(x_0), H_1(x_0, x_1), ..., H_k(x_0, x_1, ..., x_k)),$$

where for each $l = 0, ..., k$, $H_l : M_0 \times M_1 \times ... \times M_l \to M_l$ is a uniform contraction; that is, H_0 is a contraction, and for each $l = 1, .., k$, there exists $\tau_l \in [0, 1)$ such that for each $(x_0, x_1, ..., x_{l-1}) \in M_0 \times M_1 \times ... \times M_{l-1}$ and each $x_l, \widehat{x}_l \in M_l$,

$$d_l\left(H_l\left(x_0, x_1, ..., x_{l-1}, x_l\right), H_l\left(x_0, x_1, ..., x_{l-1}, \widehat{x}_l\right)\right) \leq \tau_l d\left(x_l, \widehat{x}_l\right).$$

Then F has a unique fixed point $(\overline{x}_0, \overline{x}_1, ..., \overline{x}_k)$. If, moreover, for each $l = 1, ..., k$,

$$H_l\left(., \overline{x}_l\right) : M_0 \times M_1 \times ... \times M_{l-1} \to M_l$$

is continuous, then $(\overline{x}_0, \overline{x}_1, ..., \overline{x}_k)$ is an attracting fixed point of H.

We recall that the map $\Psi : X_{0c} \to X_{0h}$ is defined by

$$\Psi(x_c) = \Pi_h \left(I - K_2 \Phi_F\right)^{-1} (K_1 x_c)(0), \ \forall x_c \in X_{0c}.$$

We define the map $\Gamma_0 : BC^\eta(\mathbb{R}, X_{0c}) \to BC^\eta(\mathbb{R}, X_0)$ by

$$\Gamma_0(u) = (I - K_2 \Phi_F)^{-1}(u), \ \forall u \in BC^\eta(\mathbb{R}, X_{0c}).$$

For each $\delta \geq 0$, the bounded linear operator $L : BC^\delta(\mathbb{R}, X_0) \to X_{0h}$ is defined by

$$L(u) = \Pi_h u(0), \ \forall u \in BC^\delta(\mathbb{R}, X_{0c}).$$

Then we have

$$\Psi(x_c) = L\Gamma_0(K_1 x_c), \ \forall x_c \in X_{0c}$$

and

$$\Gamma_0(u) = u + K_2 \Phi_F(\Gamma_0(u)), \ \forall u \in BC^\eta(\mathbb{R}, X_{0c}).$$

So we obtain

(4.25) $$\Gamma_0 = J + K_2 \circ \Phi_F \circ (\Gamma_0),$$

where J is the continuous imbedding from $BC^\eta(\mathbb{R}, X_{0c})$ into $BC^\eta(\mathbb{R}, X_0)$.

From (4.25), we deduce that for each $u \in BC^\eta(\mathbb{R}, X_{0c})$,

$$\|\Gamma_0(u) - u\|_{BC^\eta(\mathbb{R}, X_0)} \leq \|K_2\|_{\mathcal{L}(BC^\eta(\mathbb{R}, X), BC^\eta(\mathbb{R}, X_0))} |F|_{0, X_0},$$

$$\|\Pi_h \Gamma_0(u)(t)\|_{BC^0(\mathbb{R}, X)} \leq \|K_h\|_{\mathcal{L}(BC^0(\mathbb{R}, X))} \|\Pi_h F\|_{0, X_0} = \varrho_0, \ \forall t \in \mathbb{R}.$$

For each subset $E \subset BC^\eta(\mathbb{R}, X_{0c})$, denote

$$M_{0,E} = \left\{\Theta \in C\left(E, \overline{V}^0_{\varrho_0}\right) : \sup_{u \in E} \|\Theta(u) - u\|_{BC^\eta(\mathbb{R}, X_0)} < +\infty\right\}$$

and set

$$M_0 = M_{0, BC^\eta(\mathbb{R}, X_{0c})}.$$

From the above remarks, it follows that Γ_0 (respectively $\Gamma_0 \mid_E$) must be an element of M_0 (respectively $M_{0,E}$). Since $\overline{V}^0_{\varrho_0}$ is a closed subset of $BC^\eta(\mathbb{R}, X_0)$, we know that for each subset $E \subset BC^\eta(\mathbb{R}, X_{0c})$, $M_{0,E}$ is a complete metric space endowed with the metric

$$d_{0,E}\left(\Theta, \widetilde{\Theta}\right) = \sup_{u \in E} \left\|\Theta(u) - \widetilde{\Theta}(u)\right\|_{BC^\eta(\mathbb{R}, X_0)}.$$

Set

$$d_0 = d_{0, BC^\eta(\mathbb{R}, X_{0c})}.$$

LEMMA 4.16. *Let E be a Banach space and $W \in C_b^1(V_\varrho, E)$. Let $\xi \geq \delta \geq 0$ be fixed. Define $\Phi_W : V_\varrho^\eta \to BC^\xi(\mathbb{R}, E)$, $\Phi_{DW} : V_\varrho^\eta \to BC^\xi(\mathbb{R}, \mathcal{L}(X_0, E))$, and $\Phi_W^{(1)} : V_\varrho^\eta \to \mathcal{L}\left(BC^\delta(\mathbb{R}, X_0), BC^\xi(\mathbb{R}, E)\right)$ for each $t \in \mathbb{R}$, each $u \in V_\varrho^\eta$, and each $v \in BC^\delta(\mathbb{R}, X_0)$ by*

$$\Phi_W(u)(t) := W(u(t)),$$
$$\Phi_{DW}(u)(t) := DW(u(t)),$$
$$\left(\Phi_W^{(1)}(u)(v)\right)(t) := DW(u(t))(v(t)),$$

respectively. Then we have the following:

(a) *If $\xi > 0$, then Φ_W and Φ_{DW} are continuous.*
(b) *For each $u, v \in V_\varrho^\eta$, $\Phi_W^{(1)}(u) \in \mathcal{L}\left(BC^\delta(\mathbb{R}, X_0), BC^\xi(\mathbb{R}, E)\right)$,*

$$\left\|\Phi_W^{(1)}(u) - \Phi_W^{(1)}(v)\right\|_{\mathcal{L}(BC^\delta(\mathbb{R}, X_0), BC^\xi(\mathbb{R}, E))} \leq \|\Phi_{DW}(u) - \Phi_{DW}(v)\|_{BC^{\xi-\delta}(\mathbb{R}, \mathcal{L}(X_0, E))}$$

and

$$\left\|\Phi_W^{(1)}(u)\right\|_{\mathcal{L}(BC^\delta(\mathbb{R}, X_0), BC^\xi(\mathbb{R}, E))} \leq \|\Phi_{DW}(u)\|_{BC^{\xi-\delta}(\mathbb{R}, \mathcal{L}(X_0, E))} \leq |W|_{1, V_\varrho}.$$

(c) *If $\xi > \delta$, then $\Phi_W^{(1)}$ is continuous.*
(d) *If $\xi \geq \delta \geq \eta$, we have for each $u, \widehat{u} \in V_\varrho^\eta$ that*

$$\left\|\Phi_W(u) - \Phi_W(\widehat{u}) - \Phi_W^{(1)}(\widehat{u})(u - \widehat{u})\right\|_{BC^\xi(\mathbb{R}, E)} \leq \|u - \widehat{u}\|_{BC^\delta(\mathbb{R}, X_0)} \varkappa_{\xi-\delta}(u, \widehat{u})$$

where

$$\varkappa_{\xi-\delta}(u, \widehat{u}) = \sup_{s \in [0,1]} \|\Phi_{DW}(su + (1-s)\widehat{u}) - \Phi_{DW}(\widehat{u})\|_{BC^{\xi-\delta}(\mathbb{R}, \mathcal{L}(X_0, E))},$$

and if $\xi > \delta \geq \eta$, we have (by continuity of Φ_{DW})

$$\varkappa_{\xi-\delta}(u, \widehat{u}) \to 0 \quad \text{as } \|u - \widehat{u}\|_{BC^\eta(\mathbb{R}, X_0)} \to 0.$$

PROOF. We first prove that $\Phi_W \in C_b^0\left(V_\varrho^\eta, BC^\xi(\mathbb{R}, E)\right)$. For each $u, \widehat{u} \in V_\varrho^\eta$ and each $R > 0$, we have

(4.26)
$$\|\Phi_W(u) - \Phi_W(\widehat{u})\|_{BC^\xi(\mathbb{R}, E)} = \sup_{t \in \mathbb{R}} e^{-\xi|t|} \|W(u(t)) - W(\widehat{u}(t))\|$$
$$= \max\left(\sup_{|t| \leq R} e^{-\xi|t|} \|W(u(t)) - W(\widehat{u}(t))\|, 2\|W\|_0 e^{-\xi R}\right).$$

Fix some arbitrary $\varepsilon > 0$. Let $R > 0$ be such that $2\|W\|_0 e^{-\xi R} < \varepsilon$ and denote $\Omega = \{\widehat{u}(t) : |t| \leq R\}$. Since W is continuous and Ω is compact, we can find $\delta_1 > 0$ such that

$$\|W(x) - W(\widehat{x})\| \leq \varepsilon \text{ if } \widehat{x} \in \Omega, \text{ and } \|x - \widehat{x}\| \leq \delta_1.$$

Let $\delta = e^{-\eta R} \delta_1$. If $\|u - \widehat{u}\|_{BC^\eta(\mathbb{R}, X_0)} \leq \delta$, then $\|u(t) - \widehat{u}(t)\| \leq \delta_1, \forall t \in [-R, R]$, and (4.26) implies $\|\Phi_W(u) - \Phi_W(\widehat{u})\|_{BC^\xi(\mathbb{R}, E)} \leq \varepsilon$.

We now prove that $\Phi_W^{(1)} \in C\left(V_\varrho^\eta, \mathcal{L}\left(BC^\delta(\mathbb{R}, X_0), BC^\xi(\mathbb{R}, E)\right)\right)$. From the first part of the proof, since E is an arbitrary Banach space, we deduce that Φ_{DW}

is continuous. Moreover, for each $u, \widehat{u} \in V_\varrho^\eta$ and each $v \in BC^\delta(\mathbb{R}, X_0)$,

$$\left\|\left(\Phi_W^{(1)}(u)(v)\right)\right\|_{BC^\xi(\mathbb{R}, E)} = \sup_{t \in \mathbb{R}} e^{-\xi|t|} \|DW(u(t))(v(t))\|$$
$$\leq \|\Phi_{DW}(u)\|_{BC^{\xi-\delta}(\mathbb{R}, \mathcal{L}(X_0, E))} \|v\|_{BC^\delta(\mathbb{R}, X_0)}$$

and

$$\left\|\left(\left[\Phi_W^{(1)}(u) - \Phi_W^{(1)}(\widehat{u})\right](v)\right)\right\|_{BC^\xi(\mathbb{R}, E)}$$
$$\leq \|\Phi_{DW}(u) - \Phi_{DW}(\widehat{u})\|_{BC^{\xi-\delta}(\mathbb{R}, \mathcal{L}(X_0, E))} \|v\|_{BC^\delta(\mathbb{R}, X_0)}.$$

Thus, if $\xi \geq \delta$, we have for each $u \in V_\varrho^\eta$ that

$$\Phi_W^{(1)}(u) \in \mathcal{L}\left(BC^\delta(\mathbb{R}, X_0), BC^\xi(\mathbb{R}, E)\right), \quad \forall u \in V_\varrho^\eta$$

and if $\xi > \delta$,

$$\Phi_W^{(1)} \in C\left(V_\varrho^\eta, \mathcal{L}\left(BC^\delta(\mathbb{R}, X_0), BC^\xi(\mathbb{R}, E)\right)\right), \quad \forall \mu > 0.$$

Since V_ϱ is an open and convex subset of X_0, we have the following classical formula

$$W(x) - W(y) = \int_0^1 DW(sx + (1-s)y)(x-y) \, ds, \quad \forall x, y \in V_\varrho.$$

Therefore, for each $u, \widehat{u} \in V_\varrho^\eta$,

$$\left\|\Phi_W(u) - \Phi_W(\widehat{u}) - \Phi_W^{(1)}(\widehat{u})(u - \widehat{u})\right\|_{BC^\xi(\mathbb{R}, E)}$$
$$= \sup_{t \in \mathbb{R}} e^{-\xi|t|} \|W(u(t)) - W(\widehat{u}(t)) - DW(\widehat{u}(t))(u(t) - \widehat{u}(t))\|$$
$$\leq \sup_{t \in \mathbb{R}} \sup_{s \in [0,1]} e^{-\xi|t|} \|[DW(su(t) + (1-s)\widehat{u}(t)) - DW(\widehat{u}(t))](u(t) - \widehat{u}(t))\|$$
$$\leq \|u - \widehat{u}\|_{BC^\delta(\mathbb{R}, X_0)} \sup_{s \in [0,1]} \|\Phi_{DW}(su + (1-s)\widehat{u}) - \Phi_{DW}(\widehat{u})\|_{BC^{\xi-\delta}(\mathbb{R}, \mathcal{L}(X_0, E))}.$$

The proof is complete. $\square$

The following lemma is taken from Vanderbauwhede and Iooss [**106**, Lemma 3].

LEMMA 4.17. *Let E be a Banach space and $W \in C_b^1(V_\varrho, E)$. Let Φ_W and $\Phi_W^{(1)}$ be defined as in Lemma 4.16. Let $\Theta \in C\left(BC^\eta(\mathbb{R}, X_{0c}), V_\varrho^\eta\right)$ be such that*

(a) *Θ is of class C^1 from $BC^\eta(\mathbb{R}, X_{0c})$ into $BC^{\eta+\mu}(\mathbb{R}, X_0)$ for each $\mu > 0$;*
(b) *its derivative takes the form*

$$D\Theta(u)(v) = \Theta^{(1)}(u)(v), \quad \forall u, v \in BC^\eta(\mathbb{R}, X_{0c}),$$

for some globally bounded $\Theta^{(1)} : BC^\eta(\mathbb{R}, X_{0c}) \to \mathcal{L}(BC^\eta(\mathbb{R}, X_{0c}), BC^\eta(\mathbb{R}, X_0))$.

Then $\Phi_W \circ \Theta \in C_b^0(BC^\eta(\mathbb{R}, X_{0c}), BC^\eta(\mathbb{R}, E)) \cap C^1(BC^\eta(\mathbb{R}, X_{0c}), BC^{\eta+\mu}(\mathbb{R}, E))$ for each $\mu > 0$ and

$$D(\Phi_W \circ \Theta)(u)(v) = \Phi_W^{(1)}(\Theta(u)) \Theta^{(1)}(u)(v), \forall u, v \in BC^\eta(\mathbb{R}, X_{0c}).$$

4.2. SMOOTHNESS OF CENTER MANIFOLDS

PROOF. By using Lemma 4.16, it follows that
$$\Phi_W \circ \Theta \in C_b^0 \left(BC^\eta \left(\mathbb{R}, X_{0c} \right), BC^\eta \left(\mathbb{R}, E \right) \right)$$
and
$$\Phi_W^{(1)} \left(\Theta \left(. \right) \right) \Theta^{(1)} (.) \in C \left(BC^\eta \left(\mathbb{R}, X_{0c} \right), \mathcal{L} \left(BC^\eta \left(\mathbb{R}, X_{0c} \right), BC^{\eta+\mu} \left(\mathbb{R}, E \right) \right) \right).$$
Let $u, \widehat{u} \in BC^\eta \left(\mathbb{R}, X_{0c} \right)$. By Lemma 4.16, we also have
$$\left\| \Phi_W \left(\Theta \left(u \right) \right) - \Phi_W \left(\Theta \left(\widehat{u} \right) \right) - \Phi_W^{(1)} \left(\Theta \left(\widehat{u} \right) \right) \Theta^{(1)} \left(\widehat{u} \right) \left(u - \widehat{u} \right) \right\|_{BC^{\eta+\mu}(\mathbb{R},E)}$$
$$\leq \left\| \Phi_W \left(\Theta \left(u \right) \right) - \Phi_W \left(\Theta \left(\widehat{u} \right) \right) - \Phi_W^{(1)} \left(\Theta \left(\widehat{u} \right) \right) \left(\Theta \left(u \right) - \Theta \left(\widehat{u} \right) \right) \right\|_{BC^{\eta+\mu}(\mathbb{R},E)}$$
$$+ \left\| \Phi_W^{(1)} \left(\Theta \left(\widehat{u} \right) \right) \left[\Theta \left(u \right) - \Theta \left(\widehat{u} \right) - \Theta^{(1)} \left(\widehat{u} \right) \left(u - \widehat{u} \right) \right] \right\|_{BC^{\eta+\mu}(\mathbb{R},E)}$$
$$\leq \left\| \Theta \left(u \right) - \Theta \left(\widehat{u} \right) \right\|_{BC^{\eta+\mu/2}(\mathbb{R},X_0)} \varkappa_{\mu/2} \left(\Theta \left(u \right), \Theta \left(\widehat{u} \right) \right)$$
$$+ \left\| \Phi_{DW} \left(\Theta \left(\widehat{u} \right) \right) \right\|_{BC^{\mu/2}(\mathbb{R},\mathcal{L}(X_0,E))} \left\| \Theta \left(u \right) - \Theta \left(\widehat{u} \right) - \Theta^{(1)} \left(\widehat{u} \right) \left(u - \widehat{u} \right) \right\|_{BC^{\eta+\mu/2}(\mathbb{R},X_0)}$$
and the result follows. $\square$

One may extend the previous lemma to any order $k > 1$.

LEMMA 4.18. *Let E be a Banach space and let $W \in C_b^k \left(V_\varrho, E \right)$ (for some integer $k \geq 1$). Let $l \in \{1, ..., k\}$ be an integer. Suppose $\xi \geq 0, \mu \geq 0$ are two real numbers and $\delta_1, \delta_2, ..., \delta_l \geq 0$ such that $\xi = \mu + \delta_1 + \delta_2 + ... + \delta_l$. Define*
$$\Phi_{D^{(l)}W} \left(u \right) \left(t \right) := D^{(l)} W \left(u \left(t \right) \right), \forall t \in \mathbb{R}, \forall u \in V_\varrho^\eta,$$
$$Phi_W^{(l)} \left(u \right) \left(u_1, u_2, ..., u_l \right) \left(t \right) := D^{(l)} W \left(u \left(t \right) \right) \left(u_1 \left(t \right), u_2 \left(t \right), ..., u_l \left(t \right) \right),$$
$$forall t \in \mathbb{R}, \forall u \in V_\varrho^\eta, \forall u_i \in BC^{\delta_i} \left(\mathbb{R}, X_0 \right), \text{ for } i = 1, ..., l.$$
Then we have the following:

(a) *If $\xi > 0$, then $\Phi_{D^{(l)}W} : V_\varrho^\eta \to BC^\xi \left(\mathbb{R}, \mathcal{L}^{(l)} \left(X_0, E \right) \right)$ is continuous.*
(b) *For each $u, v \in V_\varrho^\eta$, $\Phi_W^{(l)}(u) \in \mathcal{L}^{(l)}(BC^{\delta_1}(\mathbb{R}, X_0), ..., BC^{\delta_l}(\mathbb{R}, X_0); BC^\xi(\mathbb{R}, E))$,*
$$\left\| \Phi_W^{(l)} \left(u \right) - \Phi_W^{(l)} \left(v \right) \right\|_{\mathcal{L}^{(l)}\left(BC^{\delta_1}(\mathbb{R},X_0),...,BC^{\delta_l}(\mathbb{R},X_0);BC^\xi(\mathbb{R},E)\right)}$$
$$\leq \left\| \Phi_{D^{(l)}W} \left(u \right) - \Phi_{D^{(l)}W} \left(v \right) \right\|_{BC^\mu\left(\mathbb{R},\mathcal{L}^{(l)}(X_0,E)\right)}$$
and
$$\left\| \Phi_W^{(l)} \left(u \right) \right\|_{\mathcal{L}^{(l)}\left(BC^{\delta_1}(\mathbb{R},X_0),...,BC^{\delta_l}(\mathbb{R},X_0);BC^\xi(\mathbb{R},E)\right)}$$
$$\leq \left\| \Phi_{D^{(l)}W} \left(u \right) \right\|_{BC^\mu\left(\mathbb{R},\mathcal{L}^{(l)}(X_0,E)\right)} \leq |W|_{l,V_\varrho}.$$

(c) *If $\mu > 0$, then $\Phi_W^{(l)}$ is continuous.*
(d) *If $\delta_1 \geq \eta$, we have for each $u, \widehat{u} \in V_\varrho^\eta$ that*
$$\left\| \Phi_W^{(l-1)} \left(u \right) - \Phi_W^{(l-1)} \left(\widehat{u} \right) - \Phi_W^{(l)} \left(\widehat{u} \right) \left(u - \widehat{u} \right) \right\|_{\mathcal{L}^{(l-1)}\left(BC^{\delta_2}(\mathbb{R},X_0),...,BC^{\delta_l}(\mathbb{R},X_0);BC^\xi(\mathbb{R},E)\right)}$$
$$\leq \left\| u - \widehat{u} \right\|_{BC^{\delta_1}(\mathbb{R},X_0)} \varkappa_\mu^{(l)} \left(u, \widehat{u} \right),$$
where
$$\varkappa_\mu^{(l)} \left(u, \widehat{u} \right) = \sup_{s \in [0,1]} \left\| \Phi_{D^{(l)}W} \left(su + (1-s)\widehat{u} \right) - \Phi_{D^{(l)}W} \left(\widehat{u} \right) \right\|_{BC^\mu\left(\mathbb{R},\mathcal{L}^{(l)}(X_0,E)\right)},$$

and if $\mu > 0$, we have by continuity of $\Phi_{D^{(l)}W}$ that
$$\varkappa_\mu^{(l)}(u,\widehat{u}) \to 0 \text{ as } \|u - \widehat{u}\|_{BC^\eta(\mathbb{R},X_0)} \to 0.$$

PROOF. This proof is similar to that of Lemma 4.16. □

In the following lemma we use a formula for the kth-derivative of the composed map. This formula is taken from Avez [6, p. 38] which also corrects the one used in Vanderbauwhede [104, Proof of Lemma 3.11].

LEMMA 4.19. *Let E be a Banach space and let $W \in C_b^k(V_\varrho, E)$. Let Φ_W and $W^{(k)}$ be defined as above. Let $\Theta \in C\left(BC^\eta(\mathbb{R}, X_{0c}), V_\varrho^\eta\right)$ be such that*

(a) Θ *is of class C^k from $BC^\eta(\mathbb{R}, X_{0c})$ into $BC^{k\eta+\mu}(\mathbb{R}, X_0)$ for each $\mu > 0$;*
(b) *for each $l = 1, ..., k$, its derivative takes the form*

$$D^l\Theta(u)(v_1, v_2, ..., v_l) = \Theta^{(l)}(u)(v_1, v_2, ..., v_l), \forall u, v_1, v_2, ..., v_l \in BC^\eta(\mathbb{R}, X_{0c}),$$

for some globally bounded $\Theta^{(l)} : BC^\eta(\mathbb{R}, X_{0c}) \to \mathcal{L}^{(l)}(BC^\eta(\mathbb{R}, X_{0c}); BC^\eta(\mathbb{R}, X_0))$.
Then $\Phi_W \circ \Theta \in C_b^0(BC^\eta(\mathbb{R}, X_{0c}), BC^\eta(\mathbb{R}, E)) \cap C^k(BC^\eta(\mathbb{R}, X_{0c}), BC^{k\eta+\mu}(\mathbb{R}, E))$ for each $\mu > 0$. Moreover, for each $l = 1, ..., k$ and each $u, v_1, v_2, ..., v_l \in BC^\eta(\mathbb{R}, X_{0c})$,

$$D^l(\Phi_W \circ \Theta)(u)(v) = (\Phi_W \circ \Theta)^{(l)}(u)(v_1, v_2, ..., v_l)$$

for some globally bounded $(\Phi_W \circ \Theta)^{(l)} : BC^\eta(\mathbb{R}, X_{0c}) \to \mathcal{L}^{(l)}(BC^\eta(\mathbb{R}, X_{0c}); BC^\eta(\mathbb{R}, E))$. More precisely, we have for $j = 1, ..., k$ that

(i) $(\Phi_W \circ \Theta)^{(j)}(u) = \Phi_W^{(1)}(\Theta(u)) D^{(j)}\Theta(u) + \widetilde{\Phi}_{W,j}(u);$
(ii) $\widetilde{\Phi}_{W,1}(u) = 0;$
(iii) *for $j > 1$, the map $\widetilde{\Phi}_{W,j}(u)$ is a finite sum $\sum_{\lambda \in \Lambda_j} \widetilde{\Phi}_{W,\lambda,j}(u)$, where for each $\lambda \in \Lambda_j$ the map $\widetilde{\Phi}_{W,\lambda,j}(u) : BC^\eta(\mathbb{R}, X_{0c}) \to \mathcal{L}^{(j)}(BC^\eta(\mathbb{R}, X_{0c}), BC^\eta(\mathbb{R}, E))$ has the following form*

$$\widetilde{\Phi}_{W,\lambda,j}(u)(u_1, u_2, ..., u_j) = \Phi_W^{(l)}(\Theta(u)) \begin{pmatrix} D^{(r_1)}\Theta(u)\left(u_{i_1^{r_1}}, u_{i_2^{r_1}}, ..., u_{i_{r_1}^{r_1}}\right), ..., \\ D^{(r_l)}\Theta(u)\left(u_{i_1^{r_l}}, ..., u_{i_{r_l}^{r_l}}\right) \end{pmatrix}$$

with $2 \le l \le j$, $1 \le r_i \le j-1$ for $1 \le i \le l$, $r_1 + r_2 + ... + r_l = j$,

$$\{i_1^{r_m}, ..., i_{r_m}^{r_m}\} \subset \{1, ..., j\}, \forall m = 1, ..., l$$
$$\{i_1^{r_m}, ..., i_{r_m}^{r_m}\} \cap \{i_1^{r_n}, ..., i_{r_n}^{r_n}\} = \emptyset, \text{ if } m \ne n,$$
$$i_1^{r_m} \le i_2^{r_m} \le ... \le i_{r_m}^{r_m}, \forall m = 1, ..., l,$$

and each $\lambda \in \Lambda_j$ corresponds to each such a particular choice.

PROOF. This proof is similar to that of Lemma 4.17. □

PROOF OF THEOREM 4.13. **Step 1. Existence of a fixed point.** Let $k, \eta,$ and $\widehat{\eta}$ be the numbers introduced in Assumption 4.12. Let $\mu > 0$ be such that $k\eta + (2k-1)\mu = \widehat{\eta}$. We first apply Lemma 4.15. For each $j = 1, ..., k$ and each subset $E \subset BC^\eta(\mathbb{R}, X_{0c})$, define $M_{j,E}$ as the Banach space of all continuous maps $\Theta_j : E \to \mathcal{L}^{(j)}\left(BC^\eta(\mathbb{R}, X_{0c}), BC^{j\eta+(2j-1)\mu}(\mathbb{R}, X_0)\right)$ such that

$$|\Theta_j|_j = \sup_{u \in E} \|\Theta_j(u)\|_{\mathcal{L}^{(j)}\left(BC^\eta(\mathbb{R}, X_{0c}), BC^{j\eta+(2j-1)\mu}(\mathbb{R}, X_0)\right)} < +\infty.$$

For $j = 0, ..., k$, define the map $H_{j,E} : M_{0,E} \times M_{1,E} \times ... \times M_{j,E} \to M_{j,E}$ as follows:
If $j = 0$, set for each $u \in E$ that
$$H_{0,E}(\Theta_0)(u) = u + K_2 \circ \Phi_F \circ \Theta_0(u).$$
If $j = 1$, set for each $u \in E$ that
$$(4.27) \qquad H_{1,E}(\Theta_0, \Theta_1)(u)(.) = J^1 + K_2 \circ \Phi_F^{(1)}(\Theta_0(u)) \circ \Theta_1(u),$$
where J^1 is the continuous imbedding from $BC^\eta(\mathbb{R}, X_{0c})$ into $BC^{\eta+\mu}(\mathbb{R}, X_0)$.

If $k \geq 2$, set for each $j = 2, ..., k$ and each $u \in E$ that
$$(4.28) \qquad \begin{aligned} &H_{j,E}(\Theta_0, \Theta_1, ..., \Theta_j)(u) \\ &= K_2 \circ \Phi_F^{(1)}(\Theta_0(u)) \circ \Theta_j(u) + \widehat{H}_{j,E}(\Theta_0, \Theta_1, ..., \Theta_{j-1})(u), \end{aligned}$$
where
$$\widehat{H}_{j,E}(\Theta_0, \Theta_1, ..., \Theta_{j-1})(u) = \sum_{\lambda \in \Lambda_j} \widehat{H}_{\lambda,j,E}(\Theta_0, \Theta_1, ..., \Theta_{j-1})(u)$$
and
$$\widehat{H}_{\lambda,j,E}(\Theta_0, \Theta_1, ..., \Theta_{j-1})(u)(u_0, u_1, ..., u_j)$$
$$= K_2 \circ \Phi_F^{(l)}(\Theta_0(u)) \left(\Theta_{r_1}(u) \left(u_{i_1^{r_1}}, u_{i_2^{r_1}}, ..., u_{i_{r_1}^{r_1}} \right), ..., \Theta_{r_l}(u) \left(u_{i_1^{r_l}}, ..., u_{i_{r_l}^{r_l}} \right) \right)$$
with the same constraints as in Lemma 4.19 for λ, r_j, l, and $i_k^{r_j}$.

Define
$$H_j = H_{j,BC^\eta(\mathbb{R},X_{0c})} \quad \text{for each } j = 0, ..., k.$$
In the above definition one has to consider K_2 as a linear operator from
$$BC^{j\eta+(2j-1)\mu}(\mathbb{R}, X)$$
into $BC^{j\eta+(2j-1)\mu}(\mathbb{R}, X_0)$, and $\Phi_F^{(l)}(\Theta_0(u))$ as an element of
$$\mathcal{L}^{(j)}\left(BC^{r_1\eta+(2r_1-1)\mu}(\mathbb{R}, X_0), ..., BC^{r_l\eta+(2r_l-1)\mu}(\mathbb{R}, X_0); BC^{j\eta+(2j-1)\mu}(\mathbb{R}, X) \right).$$
Notice that
$$j\eta + (2j-1)\mu > \sum_{k=1}^l r_k \eta + (2r_k - 1)\mu$$
since $2 \leq l \leq j$ and $r_1 + r_2 + ... + r_l = j$. Finally, define $H : M_0 \times M_1 \times ... \times M_k \to M_0 \times M_1 \times ... \times M_k$ by
$$H(\Theta_0, \Theta_1, ..., \Theta_k) = (H_0(\Theta_0), H_1(\Theta_0, \Theta_1), ..., H_k(\Theta_0, \Theta_1, ..., \Theta_k)).$$

We now check that the conditions of Lemma 4.15 are satisfied. We have already shown that H_0 is a contraction on X_0. It follows from (4.27) and (4.28) that H_j ($1 \leq j \leq k$) is a contraction on X_j. More precisely, from Assumption 4.12 c), we have for each $j = 1, ..., k$ that
$$\sup_{u \in V_\varrho^\eta} \left\| K_2 \circ \Phi_F^{(1)}(u) \right\|_{\mathcal{L}\left(BC^{j\eta+(2j-1)\mu}(\mathbb{R},X_0), BC^{j\eta+(2j-1)\mu}(\mathbb{R},X_0)\right)}$$
$$\leq \|K_2\|_{\mathcal{L}\left(BC^{j\eta+(2j-1)\mu}(\mathbb{R},X)\right)} \sup_{u \in V_\varrho^\eta} \left\| \Phi_F^{(1)}(u) \right\|_{\mathcal{L}\left(BC^{j\eta+(2j-1)\mu}(\mathbb{R},X_0), BC^{j\eta+(2j-1)\mu}(\mathbb{R},X)\right)}$$
$$\leq \sup_{\theta \in [\eta,\widehat{\eta}]} \|K_2\|_{\mathcal{L}(BC^\theta(\mathbb{R},X))} |F|_{1,V_\varrho}$$
$$\leq \sup_{\theta \in [\eta,\widehat{\eta}]} \|K_2\|_{\mathcal{L}(BC^\theta(\mathbb{R},X))} \|F\|_{\text{Lip}(X_0,X)} < 1.$$

Thus, each H_j is a contraction. The fixed point of H_0 is Γ_0, and we denote by $\Gamma = (\Gamma_0, \Gamma_1, ..., \Gamma_k)$ the fixed point of H. Moreover, for $\mu = 0$, each H_j is still a contraction so we have for each $j = 1, ..., k$ that

$$\sup_{u \in BC^\eta(\mathbb{R}, X_{0c})} \|\Gamma_j(u)\|_{\mathcal{L}^{(j)}(BC^\eta(\mathbb{R}, X_0), BC^{j\eta}(\mathbb{R}, X_0))} < +\infty.$$

Step 2. Attractivity of the fixed point. In this part we apply Lemma 4.15 to prove that for each compact subset C of $BC^\eta(\mathbb{R}, X_{0c})$ and each $\Theta \in M_0 \times M_1 \times ... \times M_k$,

(4.29) $$\lim_{m \to +\infty} H_C^m(\Theta \mid_C) = \Gamma \mid_C.$$

Let C be a compact subset of $BC^\eta(\mathbb{R}, X_{0c})$. From the definition of H_C, it is clear that

$$\Gamma \mid_C = H_C(\Gamma \mid_C)$$

and from the step 1, it is not difficult to see that for each $j = 0, ..., k$, $H_{j,C}$ is a contraction. In order to apply Lemma 4.15, it remains to prove that for each $j = 1, ..., k$, $H_{j,C}(\Theta_{0,C}, \Theta_{1,C}, ..., \Theta_{j-1,C}, \Gamma_j \mid_C) \in M_j$ dependents continuously on $(\Theta_{0,C}, \Theta_{1,C}, ..., \Theta_{j-1,C}) \in M_{0,C} \times M_{1,C} \times ... \times M_{j-1,C}$.

We have

$$H_j(\Theta_{0,C}, \Theta_{1,C}, ..., \Theta_{j-1,C}, \Gamma^{(j)} \mid_C)(u)$$
$$= K_2 \circ \Phi_F^{(1)}(\Theta_{0,C}(u)) \circ \Gamma^{(j)}(u) + \widehat{H}_j(\Theta_{0,C}, \Theta_{1,C}, ..., \Theta_{j-1,C})(u).$$

Since $\Gamma^{(j)}(u) \in \mathcal{L}^{(j)}(BC^\eta(\mathbb{R}, X_0), BC^{j\eta}(\mathbb{R}, X_0))$ and $\Phi(u) \in V_\varrho^\eta$, we can consider $\Phi_F^{(1)}$ as a map from V_ϱ^η into $\mathcal{L}(BC^{j\eta}(\mathbb{R}, X_0), BC^{j\eta+(2j-1)\mu}(\mathbb{R}, X_0))$, and by Lemma 4.16 this map is continuous.

Indeed, let $\Theta_0, \widehat{\Theta}_0 \in M_0$ be two maps. Then we have

$$\sup_{u \in C} \left\| K_2 \circ \left[\Phi_F^{(1)}(\Theta_0(u)) - \Phi_F^{(1)}(\widehat{\Theta}_0(u)) \right] \circ \Gamma^{(j)}(u) \right\|_{\mathcal{L}^{(j)}(BC^\eta(\mathbb{R}, X_{0c}), BC^{j\eta+(2j-1)\mu}(\mathbb{R}, X_0))}$$

$$\leq \|K_2\|_{\mathcal{L}(BC^{j\eta+(2j-1)\mu}(\mathbb{R}, X))}$$
$$\cdot \sup_{u \in C} \left\| \left[\Phi_F^{(1)}(\Theta_0(u)) - \Phi_F^{(1)}(\widehat{\Theta}_0(u)) \right] \circ \Gamma^{(j)}(u) \right\|_{\mathcal{L}^{(j)}(BC^\eta(\mathbb{R}, X_{0c}), BC^{j\eta+(2j-1)\mu}(\mathbb{R}, X))}$$

$$\leq \|K_2\|_{\mathcal{L}(BC^{j\eta+(2j-1)\mu}(\mathbb{R}, X_0))} \sup_{u \in C} \left\| \Gamma^{(j)}(u) \right\|_{\mathcal{L}^{(j)}(BC^\eta(\mathbb{R}, X_{0c}), BC^{j\eta}(\mathbb{R}, X_0))}$$
$$\cdot \sup_{u \in C} \left\| \Phi_F^{(1)}(\Theta_0(u)) - \Phi_F^{(1)}(\widehat{\Theta}_0(u)) \right\|_{\mathcal{L}^{(j)}(BC^{j\eta}(\mathbb{R}, X_0), BC^{j\eta+(2j-1)\mu}(\mathbb{R}, X))}$$

and by Lemma 4.16 we have

$$\sup_{u \in C} \left\| \Phi_F^{(1)}(\Theta_0(u)) - \Phi_F^{(1)}(\widehat{\Theta}_0(u)) \right\|_{\mathcal{L}^{(j)}(BC^{j\eta}(\mathbb{R}, X_0), BC^{j\eta+(2j-1)\mu}(\mathbb{R}, X))}$$

$$\leq \sup_{u \in C} \left\| \Phi_{DF}(\Theta_0(u)) - \Phi_{DF}(\widehat{\Theta}_0(u)) \right\|_{BC^{(2j-1)\mu}(\mathbb{R}, \mathcal{L}(X_0, X))}$$

$$\leq \max \left(\begin{array}{c} \sup_{|t| \geq R} e^{-(2j-1)\mu |t|} \left\| DF(\Theta_0(u)(t)) - DF(\widehat{\Theta}_0(u)(t)) \right\|_{\mathcal{L}(X_0, X)}, \\ \sup_{|t| \leq R} e^{-(2j-1)\mu |t|} \left\| DF(\Theta_0(u)(t)) - DF(\widehat{\Theta}_0(u)(t)) \right\|_{\mathcal{L}(X_0, X)} \end{array} \right).$$

4.2. SMOOTHNESS OF CENTER MANIFOLDS

Since $\widehat{\Theta}_0$ is continuous, C is compact, it follows that $\widehat{\Theta}_0(C)$ is compact, and by Ascoli's theorem (see for example Lang [**70**]), the set $\widehat{C} = \bigcup_{|t| \leq R, u \in C} \left\{ \widehat{\Theta}_0(u)(t) \right\}$ is compact. But since $DF(.)$ is continuous, we have that for each $\varepsilon > 0$, there exists $\eta > 0$, such that for each $x, y \in X_0$,

$$d\left(x, \widehat{C}\right) \leq \eta, \quad d\left(y, \widehat{C}\right) \leq \eta, \quad \text{and} \quad \|x - y\| \leq \eta \Rightarrow \|DF(x) - DF(y)\| \leq \varepsilon.$$

Hence, the map $\Theta_{0,C} \to K_2 \circ \Phi_F^{(1)}(\Theta_{0,C}(.)) \circ \Gamma^{(j)}(.)$ is continuous.

It remains to consider $1 \leq r_i \leq j - 1$, $r_1 + r_2 + \ldots + r_l = j$. We have

$$\left\| K_2 \circ \left[\Phi_F^{(l)}(\Theta_0(u)) - \Phi_F^{(l)}\left(\widehat{\Theta}_0(u)\right) \right] \left(\widetilde{\Theta}_{r_1}(u), \ldots, \widetilde{\Theta}_{r_l}(u) \right) \right\|_{\mathcal{L}^{(j)}\left(BC^{\eta}(\mathbb{R}, X_{0c}), BC^{j\eta + (2j-1)\mu}(\mathbb{R}, X_0)\right)}$$

$$\leq \|K_2\|_{\mathcal{L}\left(BC^{j\eta + (2j-1)\mu}(\mathbb{R}, X), BC^{j\eta + (2j-1)\mu}(\mathbb{R}, X_0)\right)}$$

$$\cdot \sup_{u \in C} \left\| \left[\Phi_F^{(l)}(\Theta_0(u)) - \Phi_F^{(l)}\left(\widehat{\Theta}_0(u)\right) \right] \left(\widetilde{\Theta}_{r_1}(u), \ldots, \widetilde{\Theta}_{r_l}(u) \right) \right\|_{\mathcal{L}^{(j)}\left(BC^{\eta}(\mathbb{R}, X_{0c}), BC^{j\eta + (2j-1)\mu}(\mathbb{R}, X)\right)}$$

$$\leq \|K_2\|_{\mathcal{L}\left(BC^{j\eta + (2j-1)\mu}(\mathbb{R}, X), BC^{j\eta + (2j-1)\mu}(\mathbb{R}, X_0)\right)}$$

$$\cdot \left\| \Phi_F^{(l)}(\Theta_0(u)) - \Phi_F^{(l)}\left(\widehat{\Theta}_0(u)\right) \right\|_{\mathcal{L}^{(l)}\left(\prod_{p=1,\ldots,l} BC^{r_p\eta + (2r_p-1)\mu}(\mathbb{R}, X_0); BC^{j\eta + (2j-1)\mu}(\mathbb{R}, X)\right)}$$

$$\cdot \prod_{p=1,\ldots,l} \left\| \widetilde{\Theta}_{r_p}(u) \right\|_{\mathcal{L}^{(j)}\left(BC^{\eta}(\mathbb{R}, X_{0c}), BC^{r_p\eta + (2r_p-1)\mu}(\mathbb{R}, X_0)\right)}$$

and by Lemma 4.18 we have

$$\sup_{u \in C} \left\| \Phi_F^{(l)}(\Theta_0(u)) - \Phi_F^{(l)}\left(\widehat{\Theta}_0(u)\right) \right\|_{\mathcal{L}^{(l)}\left(\prod_{p=1,\ldots,l} BC^{r_p\eta + (2r_p-1)\mu}(\mathbb{R}, X_0); BC^{j\eta + (2j-1)\mu}(\mathbb{R}, X)\right)}$$

$$\leq \sup_{u \in C} \left\| \Phi_{D^{(l)}F}(\Theta_0(u)) - \Phi_{D^{(l)}F}\left(\widehat{\Theta}_0(u)\right) \right\|_{BC^{\delta}\left(\mathbb{R}, \mathcal{L}^{(l)}(X_0, X)\right)}$$

with $\delta = (j\eta + (2j-1)\mu) - \sum_{k=1}^{l} r_k \eta + (2r_k - 1)\mu > 0$. By using the same compactness arguments as previously, we deduce that

$$\sup_{u \in C} \left\| \Phi_{D^{(l)}F}(\Theta_0(u)) - \Phi_{D^{(l)}F}\left(\widehat{\Theta}_0(u)\right) \right\|_{BC^{\delta}\left(\mathbb{R}, \mathcal{L}^{(l)}(X_0, X)\right)} \to 0$$

as $d_{0,C}(\Theta_0, \widehat{\Theta}_0) \to 0$. We conclude that the continuity condition of Lemma 4.15 is satisfied for each $H_{j,C}$ and (4.29) follows.

Step 3. In order to prove Theorem 4.13 it now remains to prove that for each $u, v \in BC^{\eta}(\mathbb{R}, X_{0c}), \forall j = 1, \ldots, k$,

$$(4.30) \qquad \Gamma_{j-1}(u) - \Gamma_{j-1}(v) = \int_0^1 \Gamma_j(s(u-v) + v)(u-v) \, ds,$$

where the last integral is a Riemann integral. As assumed that (4.30) is satisfied, we deduce that $\Gamma_0 : BC^{\eta}(\mathbb{R}, X_{0c}) \to BC^{k\eta + (2k-1)\mu}(\mathbb{R}, X_0)$ is k-times continuously

differentiable, and since
$$\Psi(x_c) = L \circ \Gamma_0 \circ K_1(x_c)$$
and L is a bounded linear operator from $BC^{k\eta+(2k-1)\mu}(\mathbb{R}, X_0)$ into X_{0h}, we know that $\Psi: X_{0c} \to X_{0h}$ is k-times continuously differentiable.

We now prove (4.30). Set
$$\Theta^0 = (\Theta_0^0, \Theta_1^0, ..., \Theta_k^0)$$
with
$$\Theta_0^0(u) = u, \Theta_1^0(u) = J, \text{ and } \Theta_j^0 = 0, \forall j = 2, ..., k$$
and set
$$\Theta^m = (\Theta_0^m, \Theta_1^m, ..., \Theta_k^m) = H^m(\Theta^0), \forall m \geq 1.$$
Then from Lemma 4.19, we know that $\Theta_0^m : BC^\eta(\mathbb{R}, X_{0c}) \to BC^{k\eta+(2k-1)\mu}(\mathbb{R}, X_0)$ is a C^k-map and
$$D^j \Theta_0^m(u) = \Theta_j^m(u), \quad \forall j = 1, ..., k, \quad \forall u \in BC^\eta(\mathbb{R}, X_{0c}).$$
For each $u, v \in BC^\eta(\mathbb{R}, X_{0c})$ and each $\forall j = 1, ..., k, \forall m \geq 1$,
$$\Theta_{j-1}^m(u) - \Theta_{j-1}^m(v) = \int_0^1 \Theta_j^m(s(u-v)+v)(u-v)\,ds.$$
Let $u, v \in BC^\eta(\mathbb{R}, X_{0c})$ be fixed. Denote
$$C = \{s(u-v)+v : s \in [0,1]\}.$$
Then clearly C is a compact set, and from step 2, we have for each $j = 0, ..., k$ that
$$\sup_{w \in C} \|\Theta_j^m(w) - \Gamma_j(w)\|_{BC^{j\eta+(2j-1)\mu}(\mathbb{R}, X_0)} \to 0 \text{ as } m \to +\infty.$$
Thus, (4.30) follows. $\square$

It follows from the foregoing treatment that we can obtain the derivatives of $\Gamma_0(u)$ at $u = 0$. Assume that $F(0) = 0$ and $DF(0) = 0$, we have

(4.31)
$$\begin{aligned}
&D\Gamma_0(0) = J, \\
&D^{(2)}\Gamma_0(0)(u_1, u_2) = K_2 \circ \Phi_F^{(2)}(0)\left(D\Gamma_0(0)(u_1), D\Gamma_0(0)(u_2)\right), \\
&D^{(3)}\Gamma_0(0)(u_1, u_2, u_3) = K_2 \circ \Phi_F^{(2)}(0)\left(D^{(2)}\Gamma_0(0)(u_1, u_3), D\Gamma_0(0)(u_2)\right) \\
&\quad + K_2 \circ \Phi_F^{(2)}(0)\left(D\Gamma_0(0)(u_1), D^{(2)}\Gamma_0(0)(u_2, u_3)\right) \\
&\quad + K_2 \circ \Phi_F^{(3)}(0)\left(D\Gamma_0(0)(u_1), D\Gamma_0(0)(u_2), D\Gamma_0(0)(u_3)\right), \\
&\vdots \\
&D^{(l)}\Gamma_0(0) = \sum_{\lambda \in \Lambda_j} K_2 \circ \Phi_F^{(l)}(0)\left(D^{(r_1)}\Gamma(0), ..., D\Gamma^{(r_l)}(0)\right).
\end{aligned}$$

We have the following Lemma.

LEMMA 4.20. *Let Assumptions 4.1 and 4.12 be satisfied. Assume also that $F(0) = 0$ and $DF(0) = 0$. Then*
$$\Psi(0) = 0 \quad D\Psi(0) = 0,$$
and if $k > 1$,
$$D^j \Psi(0)(x_1, ..., x_n) = \Pi_h D^{(l)} \Gamma_0(0)(K_1 x_1, ..., K_1 x_n)(0),$$
where $D^{(l)}\Gamma_0(0)$ is given by (4.31). In particular, if $k > 1$ and
$$\Pi_h D^j F(0)|_{X_{0c} \times ... \times X_{0c}} = 0 \text{ for } 2 \leq j \leq k,$$

then
$$D^j \Psi(0) = 0 \quad for \ 1 \leq j \leq k.$$

In the context of Hopf bifurcation, we need an explicit formula for $D^2\Psi(0)$. Since $D\Gamma_0(0) = J$, we obtain from the above formula that $\forall x_1, x_2 \in X_{0c}$,

$$D^2 \Psi(0)(x_1, x_2) = \Pi_h K_h \left[D^{(2)} F(0)(K_1 x_1, K_1 x_2) \right](0),$$

where

$$K_h = K_s + K_u, \quad K_1(x_c)(t) := e^{A_{0c}t} x_c,$$
$$K_u(f)(t) := -\int_t^{+\infty} e^{-A_{0u}(l-t)} \Pi_u f(l) dl,$$

and

$$K_s(f)(t) := \lim_{r \to -\infty} \Pi_{0s} \left(S_A \diamond f(r + .) \right)(t - r).$$

Hence,

$$D^2 \Psi(0)(x_1, x_2)$$
$$= -\int_0^{+\infty} e^{-A_{0u}l} \Pi_u D^{(2)} F(0) \left(e^{A_{0c}l} x_1, e^{A_{0c}l} x_2 \right) dl$$
$$+ \lim_{r \to -\infty} \Pi_{0s} \left(S_A \diamond D^{(2)} F(0) \left(e^{A_{0c}(r+.)} x_1, e^{A_{0c}(r+.)} x_2 \right) \right)(-r).$$

In order to explicit the term of the above formula, we remark that

$$(\lambda I - A)^{-1} \lim_{r \to -\infty} \Pi_{0s} \left(S_A \diamond D^{(2)} F(0) \left(e^{A_{0c}(r+.)} x_1, e^{A_{0c}(r+.)} x_2 \right) \right)(-r)$$
$$= \lim_{r \to -\infty} \Pi_{0s} \int_0^{-r} T_{A_0}(-r - s)(\lambda I - A)^{-1} D^{(2)} F(0) \left(e^{A_{0c}(r+s)} x_1, e^{A_{0c}(r+s)} x_2 \right) ds$$
$$= \lim_{r \to -\infty} \int_0^{-r} T_{A_0}(l)(\lambda I - A)^{-1} D^{(2)} F(0) \left(e^{-A_{0c}l} x_1, e^{-A_{0c}l} x_2 \right) dl$$
$$= \int_0^{+\infty} T_{A_0}(l) \Pi_{0s} (\lambda I - A)^{-1} D^{(2)} F(0) \left(e^{-A_{0c}l} x_1, e^{-A_{0c}l} x_2 \right) dl.$$

Therefore, we obtain the following formula

$$D^2 \Psi(0)(x_1, x_2)$$
$$= -\int_0^{+\infty} e^{-A_{0u}l} \Pi_u D^{(2)} F(0) \left(e^{A_{0c}l} x_1, e^{A_{0c}l} x_2 \right) dl$$
$$+ \lim_{\lambda \to +\infty} \int_0^{+\infty} T_{A_0}(l) \Pi_{0s} \lambda (\lambda I - A)^{-1} D^{(2)} F(0) \left(e^{-A_{0c}l} x_1, e^{-A_{0c}l} x_2 \right) dl.$$

Assume that X is a complex Banach space and F is twice continuously differentiable in X considered as a $\mathbb{C}$-Banach space. We assume in addition that A_{0c} is diagonalizable, and denote by $\{v_1, ..., v_n\}$ a basis of X_c such that for each $i = 1, ..., n$,

$A_{0c}v_i = \lambda_i v_i$. Then by Assumption 4.1, we must have $\lambda_i \in i\mathbb{R}, \forall i = 1, ..., n$. Moreover, for each $i, j = 1, ..., n$, we have

$$D^2\Psi(0)(v_i, v_j)$$
$$= -\int_0^{+\infty} e^{(\lambda_i+\lambda_j)l}e^{-A_{0u}l}\Pi_u D^{(2)}F(0)(v_i, v_j) dl$$
$$+ \lim_{\lambda \to +\infty} \int_0^{+\infty} T_{A_0}(l)\Pi_{0s}\lambda(\lambda I - A)^{-1}D^{(2)}F(0)\left(e^{-\lambda_i l}v_i, e^{-\lambda_j l}v_j\right) dl$$
$$= -\left(-(\lambda_i + \lambda_j)I - (-A_{0u})\right)^{-1}\Pi_u D^{(2)}F(0)(v_i, v_j)$$
$$+ \lim_{\lambda \to +\infty} \int_0^{+\infty} e^{-(\lambda_i+\lambda_j)l}T_{A_{0,s}}(l)\Pi_{0s}\lambda(\lambda I - A)^{-1}D^{(2)}F(0)(v_i, v_j) dl$$
$$= -\left(-(\lambda_i + \lambda_j)I - (-A_{0u})\right)^{-1}\Pi_u D^{(2)}F(0)(v_i, v_j)$$
$$+ \lim_{\lambda \to +\infty} \lambda(\lambda I - A)^{-1}\left((\lambda_i + \lambda_j)I - A_s\right)^{-1}\Pi_s D^{(2)}F(0)(v_i, v_j).$$

Thus,

$$\begin{aligned} D^2\Psi(0)(v_i, v_j) &= \left((\lambda_i + \lambda_j)I - A_{0u}\right)^{-1}\Pi_u D^{(2)}F(0)(v_i, v_j) \\ &+ \left((\lambda_i + \lambda_j)I - A_s\right)^{-1}\Pi_s D^{(2)}F(0)(v_i, v_j). \end{aligned}$$

Note that by Assumption 4.1 $i\mathbb{R} \subset \rho(A_s)$, so the above formula is well defined.

As in Vanderbauwhede and Iooss [106, Theorem 3], we have the following theorem about the existence of the local center manifold.

THEOREM 4.21. *Let Assumption 4.1 be satisfied. Let $F : B_{X_0}(0, \varepsilon) \to X$ be a map. Assume there exists an integer $k \geq 1$ such that F is k-time continuously differentiable in some neighborhood of 0 with $F(0) = 0$ and $DF(0) = 0$. Then there exist a neighborhood Ω of the origin in X_0 and a map $\Psi \in C_b^k(X_{0c}, X_{0h})$, with $\Psi(0) = 0$ and $D\Psi(0) = 0$, such that the following properties hold:*

(i) *If I is an interval of $\mathbb{R}$ and $x_c : I \to X_{0c}$ is a solution of*

(4.32)
$$\frac{dx_c(t)}{dt} = A_{0c}x_c(t) + \Pi_c F[x_c(t) + \Psi(x_c(t))]$$

such that

$$u(t) := x_c(t) + \Psi(x_c(t)) \in \Omega, \forall t \in I,$$

then for each $t, s \in I$ with $t \geq s$,

$$u(t) = u(s) + A\int_s^t u(l)dl + \int_s^t F(u(l)) dl.$$

(ii) *If $u : \mathbb{R} \to X_0$ is a map such that for each $t, s \in \mathbb{R}$ with $t \geq s$,*

$$u(t) = u(s) + A\int_s^t u(l)dl + \int_s^t F(u(l)) dl$$

and $u(t) \in \Omega$, $\forall t \in \mathbb{R}$, then

$$\Pi_h u(t) = \Psi(\Pi_c u(t)), \forall t \in \mathbb{R},$$

and $\Pi_c u : \mathbb{R} \to X_{0c}$ is a solution of (4.32).

(iii) If $k \geq 2$, then for each $x_1, x_2 \in X_{0c}$,

$$D^2\Psi(0)(x_1, x_2)$$
$$= -\int_0^{+\infty} e^{-A_{0u}l}\Pi_u D^{(2)}F(0)\left(e^{A_{0c}l}x_1, e^{A_{0c}l}x_2\right) dl$$
$$+ \lim_{r \to -\infty} \Pi_{0s}\left(S_A \diamond D^{(2)}F(0)\left(e^{A_{0c}(r+\cdot)}x_1, e^{A_{0c}(r+\cdot)}x_2\right)\right)(-r).$$

Moreover, X is a $\mathbb{C}$-Banach space, and if $\{v_1, ..., v_n\}$ is a basis of X_c such that for each $i = 1, ..., n$, $A_{0c}v_i = \lambda_i v_i$, with $\lambda_i \in i\mathbb{R}$, then for each $i, j = 1, ..., n$,

$$\begin{aligned}D^2\Psi(0)(v_i, v_j) &= ((\lambda_i + \lambda_j)I - A_{0u})^{-1}\Pi_u D^{(2)}F(0)(v_i, v_j) \\ &+ ((\lambda_i + \lambda_j)I - A_s)^{-1}\Pi_s D^{(2)}F(0)(v_i, v_j).\end{aligned}$$

PROOF. Set for each $r > 0$ that
$$F_r(x) = F(x)\chi_c\left(r^{-1}\Pi_{0c}(x)\right)\chi_h\left(r^{-1}\|\Pi_{0h}(x)\|\right), \forall x \in X_0,$$
where $\chi_c : X_{0c} \to [0, +\infty)$ is a C^∞ map with $\chi_{0c}(x) = 1$ if $\|x\| \leq 1$, $\chi_{0c}(x) = 0$ if $\|x\| \geq 2$, and $\chi_h : [0, +\infty) \to [0, +\infty)$ is a C^∞ map with $\chi_h(x) = 1$ if $|x| \leq 1$, $\chi_h(x) = 0$ if $|x| \geq 2$. Then by using the same arguments as in the proof of Theorem 3 in [**106**], we deduce that there exists $r_0 > 0$, such that for each $r \in (0, r_0]$, F_r satisfies Assumption 4.12. By applying Theorem 4.13 to
$$\frac{du(t)}{dt} = Au(t) + F_r(u(t)), \ t \geq 0, \text{ and } u(0) = x \in \overline{D(A)}$$
for $r > 0$ small enough, the result follows. □

In order to investigate the existence of Hopf bifurcation we also need the following result.

PROPOSITION 4.22. *Let the assumptions of Theorem 4.21 be satisfied. Assume that $\overline{x} \in X_0$ is an equilibrium of $\{U(t)\}_{t \geq 0}$ (i.e. $\overline{x} \in D(A)$ and $A\overline{x} + F(\overline{x}) = 0$) such that*
$$\overline{x} \in \Omega.$$
Then
$$\Pi_{0h}\overline{x} = \Psi(\Pi_{0c}\overline{x})$$
and $\Pi_{0c}\overline{x}$ is an equilibrium of the reduced equation (4.32). Moreover, if we consider the linearized equation (4.32) at $\Pi_{0c}\overline{x}$
$$\frac{dy_c(t)}{dt} = L(\overline{x})y_c(t)$$
with
$$L(\overline{x}) = [A_{0c} + \Pi_c DF(\overline{x})[I + D\Psi(\Pi_{0c}\overline{x})]],$$
then we have the following spectral properties
$$\sigma(L(\overline{x})) = \sigma((A + DF(\overline{x}))_0) \cap \{\lambda \in \mathbb{C} : \operatorname{Re}(\lambda) \in [-\eta, \eta]\}.$$

PROOF. Let $\overline{x} \in X_0$ be an equilibrium of $\{U(t)\}_{t \geq 0}$ such that $\overline{x} \in \Omega$. We set
$$\overline{x}_c = \Pi_c \overline{x} \text{ and } \overline{u}(t) = \overline{x}, \forall t \in \mathbb{R}.$$
Then the linearized equation at $\overline{x}$ is given by

(4.33) $$\frac{dw(t)}{dt} = (A + DF(\overline{x}))w(t), \text{ for } t \geq 0, \text{ and } w(0) = w_0 \in X_0.$$

So
$$w(t) = T_{(A+DF(\overline{x}))_0}(t)w_0, \forall t \geq 0.$$
Moreover, we have
$$D\Psi(x_c) y_c = \Pi_h \left[\Gamma_0^1(\overline{u}) (K_1 y_c) \right]$$
and
$$\Gamma_0^1(\overline{u})(v) = v + K_2 \Phi_{DF(\overline{x})} \left(\Gamma_0^1(u)(v) \right), \forall v \in BC^\eta (\mathbb{R}, X_{0c}).$$
It follows that
$$\Gamma_0^1(\overline{u}) = \left(I - K_2 \Phi_{DF(\overline{x})} \right)^{-1} v.$$
Thus,
$$D\Psi(\overline{x}_c) y_c = \Pi_h \left[\left(I - K_2 \Phi_{DF(\overline{x})} \right)^{-1} (K_1 y_c) \right].$$
This is exactly the formula for the center manifold of equation (4.32) (see (4.23) in the proof of Theorem 4.10). By applying Theorem 4.10 to equation (4.33), we deduce that
$$W_\eta = \{ y_c + D\Psi(\overline{x}_c) y_c : y_c \in X_{0c} \}$$
is invariant by $\left\{ T_{(A+DF(\overline{x}))_0}(t) \right\}_{t \geq 0}$. Moreover, for each $w \in C(\mathbb{R}, X_0)$ the following statements are equivalent:

(1) $w \in BC^\eta(\mathbb{R}, X_0)$ is a complete orbit of $\left\{ T_{(A+DF(\overline{x}))_0}(t) \right\}_{t \geq 0}$.

(2) $\Pi_{0h} w(t) = D\Psi(\overline{x}_c)(\Pi_{0c} w(t)), \forall t \in \mathbb{R}$, and $\Pi_{0c} w(.) : \mathbb{R} \to \overline{X}_{0c}$ is a solution of the ordinary differential equation
$$\frac{dw_c(t)}{dt} = A_{0c} w_c(t) + \Pi_c DF(\overline{x}) \left[w_c(t) + D\Psi(\overline{x}_c)(w_c(t)) \right].$$
The result follows from the above equivalence. $\square$

CHAPTER 5

Hopf Bifurcation in Age Structured Models

In order to illustrate Theorem 4.21, we consider an age-structured model. Let $u(t,a)$ denote the density of a population at time t with age a. Consider the following age structured model

(5.1)
$$\begin{cases} \dfrac{\partial u(t,a)}{\partial t} + \dfrac{\partial u(t,a)}{\partial a} = -\mu u(t,a),\ a \in (0,+\infty), \\ u(t,0) = \alpha h\left(\int_0^{+\infty} \gamma(a)u(t,a)da\right), \\ u(0,.) = \varphi \in L_+^1\left((0,+\infty);\mathbb{R}\right), \end{cases}$$

where $\mu > 0$ is the mortality rate of the population, the function $h(\cdot)$ describes the fertility of the population, $\alpha \geq 0$ is considered as a bifurcation parameter.

Age structured models have been studied extensively by many researchers (Hoppensteadt [**57**], Webb [**108**], Iannelli [**59**], and Cushing [**27**]). The existence of non-trivial periodic solutions induced by Hopf bifurcation has been observed in various specific age structured models (Cushing [**25, 26**], Prüss [**89**], Swart [**96**], Kostava and Li [**67**], Bertoni [**10**]). However, there is no general Hopf bifurcation theorem that can be applied to age structured models. In this chapter, we shall use the center manifold theorem (Theorem 4.21) to establish a Hopf bifurcation theorem for the age structured model (5.1); namely, we will prove that a Hopf bifurcation occurs in the age structured model (5.1), thus a non-trivial periodic solution bifurcates from the equilibrium of (5.1) when the bifurcation parameter takes some critical values.

We first make an assumption on the fertility function $h(\cdot)$.

ASSUMPTION 5.1. *Assume that $h : \mathbb{R} \to \mathbb{R}$ is defined by*
$$h(x) = x\exp(-\beta x),\ \forall x \in \mathbb{R},$$
where $\beta > 0$ and $\gamma \in L_+^\infty\left((0,+\infty),\mathbb{R}\right)$ with
$$\int_0^{+\infty} \gamma(a)e^{-\mu a}da = 1.$$

Set
$$Y = \mathbb{R} \times L^1\left((0,+\infty);\mathbb{R}\right), \qquad Y_0 = \{0\} \times L^1\left((0,+\infty);\mathbb{R}\right),$$
$$Y_+ = \mathbb{R}_+ \times L^1\left((0,+\infty);\mathbb{R}_+\right), \quad Y_{0+} = Y_0 \cap Y_+.$$

Assume that Y is endowed with the product norm
$$\|x\| = |\alpha| + \|\varphi\|_{L^1((0,+\infty);\mathbb{R})},\ \forall x = \begin{pmatrix} \alpha \\ \varphi \end{pmatrix} \in Y.$$

We denote by
$$Y^\mathbb{C} = Y + iY \text{ and } Y_0^\mathbb{C} = Y_0 + iY_0$$

the complexified Banach space of Y and Y_0, respectively. We can identify $Y^{\mathbb{C}}$ to
$$Y = \mathbb{C} \times L^1((0, +\infty); \mathbb{C})$$
endowed with the product norm
$$\|x\| = |\alpha| + \|\varphi\|_{L^1((0,+\infty);\mathbb{C})}, \quad \forall x = \begin{pmatrix} \alpha \\ \varphi \end{pmatrix} \in Y^{\mathbb{C}}.$$
From now on, for each $x \in Y$, we denote by
$$\overline{x} = \begin{pmatrix} \overline{\alpha} \\ \overline{\varphi} \end{pmatrix}, \quad \operatorname{Re}(x) = \frac{x + \overline{x}}{2}, \quad \text{and} \quad \operatorname{Im}(x) = \frac{x - \overline{x}}{2}.$$
We consider the linear operator $\widehat{A} : D(\widehat{A}) \subset Y \to Y$ defined by
$$\widehat{A}\begin{pmatrix} 0 \\ \varphi \end{pmatrix} = \begin{pmatrix} -\varphi(0) \\ -\varphi' - \mu\varphi \end{pmatrix}$$
with
$$D(\widehat{A}) = \{0\} \times W^{1,1}((0, +\infty); \mathbb{R}).$$
Moreover, for each $\lambda \in \mathbb{C}$ with $\operatorname{Re}(\lambda) > -\mu$, we have $\lambda \in \rho\left(\widehat{A}\right)$ and
$$\left(\lambda I - \widehat{A}\right)^{-1} \begin{pmatrix} \alpha \\ \psi \end{pmatrix} = \begin{pmatrix} 0 \\ \varphi \end{pmatrix} \Leftrightarrow \varphi(a) = e^{-(\lambda+\mu)a}\alpha + \int_0^a e^{-(\lambda+\mu)(a-s)}\psi(s)ds.$$
Note that
$$\lambda \in \rho\left(\widehat{A}\right) \Leftrightarrow \overline{\lambda} \in \rho\left(\widehat{A}\right)$$
and
$$\left(\lambda I - \widehat{A}\right)^{-1} x = \overline{\left(\overline{\lambda} I - \widehat{A}\right)^{-1} \overline{x}}, \quad \forall x \in Y, \ \forall \lambda \in \rho\left(\widehat{A}\right).$$
It is well known that $\widehat{A}$ is a Hille-Yosida operator. Moreover, $\widehat{A}_0$ is the part of $\widehat{A}$ in Y_0 generated a C_0-semigroup of bounded linear operators $\left\{T_{\widehat{A}_0}(t)\right\}_{t \geq 0}$, which is defined by
$$T_{\widehat{A}_0}(t)\begin{pmatrix} 0 \\ \varphi \end{pmatrix} = \begin{pmatrix} 0 \\ \widehat{T}_{\widehat{A}_0}(t)\varphi \end{pmatrix},$$
where
$$\widehat{T}_{\widehat{A}_0}(t)(\varphi)(a) = \begin{cases} e^{-\mu t}\varphi(a - t), & \text{if } a \geq t, \\ 0, & \text{if } a \leq t. \end{cases}$$
$\left\{S_{\widehat{A}}(t)\right\}_{t \geq 0}$ is the integrated semigroup generated by $\widehat{A}$ and is defined by
$$S_{\widehat{A}}(t)\begin{pmatrix} \alpha \\ \varphi \end{pmatrix} = \begin{pmatrix} 0 \\ L(t)\alpha + \int_0^t \widehat{T}_{\widehat{A}_0}(s)\varphi \, ds \end{pmatrix},$$
where
$$L(t)(\alpha)(a) = \begin{cases} 0, & \text{if } a \geq t, \\ e^{-\mu a}\alpha, & \text{if } a \leq t. \end{cases}$$
Define $H : Y_0 \to Y$ and $H_1 : Y_0 \to \mathbb{R}$ by
$$H\begin{pmatrix} 0 \\ \varphi \end{pmatrix} = \begin{pmatrix} H_1\begin{pmatrix} 0 \\ \varphi \end{pmatrix} \\ 0 \end{pmatrix}, \quad H_1\begin{pmatrix} 0 \\ \varphi \end{pmatrix} = h\left(\int_0^{+\infty} \gamma(a)\varphi(a)da\right).$$

Then by identifying $u(t)$ to $v(t) = \begin{pmatrix} 0 \\ u(t) \end{pmatrix}$ the problem (5.1) can be considered as the following Cauchy problem

(5.2) $$\frac{dv(t)}{dt} = \widehat{A}v(t) + \alpha H(v(t)) \text{ for } t \geq 0, \ v(t) = y \in Y_{0+}.$$

Since h is Lipschitz continuous on $[0, +\infty)$, the following lemma is a consequence of the results in Thieme [**99**].

LEMMA 5.2. *Let Assumption 5.1 be satisfied. Then for each $\alpha \geq 0$, there exists a family of continuous maps $\{U_\alpha(t)\}_{t \geq 0}$ on Y_{0+} such that for each $y \in Y_{0+}$, the map $t \to U_\alpha(t)y$ is the unique integrated solution of (5.2), that is,*

$$U_\alpha(t)y = y + \widehat{A}\int_0^t U_\alpha(s)y ds + \int_0^t \alpha H(U_\alpha(l)y)dl, \ \forall t \geq 0,$$

or equivalently

$$U_\alpha(t)y = T_{\widehat{A}_0}(t)y + \frac{d}{dt}\left(S_{\widehat{A}} * \alpha H(U_\alpha(.)y)\right)(t), \ \forall t \geq 0.$$

Moreover, $\{U_\alpha(t)\}_{t \geq 0}$ is a continuous semiflow, that is, $U(0) = Id$,

$$U_\alpha(t)U_\alpha(s) = U_\alpha(t+s), \forall t, s \geq 0,$$

and the map $(t, x) \to U_\alpha(t)x$ is continuous from $[0, +\infty) \times Y_{0+}$ into Y_{0+}.

We recall that $\overline{y} \in Y_{0+}$ is an equilibrium of $\{U_\alpha(t)\}_{t \geq 0}$ if and only if

$$\overline{y} \in D(\widehat{A}) \text{ and } \widehat{A}\overline{y} + \alpha H(\overline{y}) = 0.$$

Here if $\alpha > 1$, equation (5.1) has two non-negative equilibria given by

$$\overline{v} = \begin{pmatrix} 0 \\ \overline{u} \end{pmatrix} \text{ with } \overline{u}(a) = Ce^{-\mu a},$$

where C is a solution of

$$C = \alpha h\left(C \int_0^{+\infty} \gamma(a)e^{-\mu a}da\right) \text{ with } C \geq 0.$$

But by Assumption 5.1 we have $\int_0^{+\infty} \gamma(a)e^{-\mu a}da = 1$, so we obtain

$$C = 0 \text{ or } C = \overline{C}(\alpha) := \beta^{-1}\ln(\alpha).$$

From now on we set

(5.3) $$\overline{v}_\alpha = \begin{pmatrix} 0 \\ \overline{u}_\alpha \end{pmatrix} \text{ with } \overline{u}(a) = \overline{C}(\alpha)e^{-\mu a}, \ \forall \alpha > 1.$$

We have

$$\alpha H(\overline{v}_\alpha) = \begin{pmatrix} \overline{C}(\alpha) \\ 0 \end{pmatrix},$$

$$\alpha DH(\psi)\begin{pmatrix} 0 \\ \varphi \end{pmatrix} = \begin{pmatrix} \alpha h'\left(\int_0^{+\infty} \gamma(a)\psi(a)da\right)\int_0^{+\infty} \gamma(a)\varphi(a)da \\ 0 \end{pmatrix},$$

so

$$\alpha DH(\overline{v}_\alpha)\begin{pmatrix} 0 \\ \varphi \end{pmatrix} = \begin{pmatrix} \eta(\alpha)\int_0^{+\infty} \gamma(a)\varphi(a)da \\ 0 \end{pmatrix},$$

where

$$\eta(\alpha) = \alpha h'\left(\int_0^{+\infty} \gamma(a)e^{-\mu a}da \overline{C}(\alpha)\right) = \alpha\left(1 - \beta\overline{C}(\alpha)\right)\exp\left(-\beta\overline{C}(\alpha)\right) = 1 - \ln(\alpha).$$

We also have for $k \geq 1$ that

$$\alpha D^k H(\psi)\left(\begin{pmatrix} 0 \\ \varphi_1 \end{pmatrix}, ..., \begin{pmatrix} 0 \\ \varphi_k \end{pmatrix}\right)$$
$$= \begin{pmatrix} \alpha h^{(k)}\left(\int_0^{+\infty} \gamma(a)\psi(a)da\right) \prod_{i=1}^k \int_0^{+\infty} \gamma(a)\varphi_i(a)da \\ 0 \end{pmatrix}.$$

The characteristic equation of the problem is

(5.4) $\quad 1 = \eta(\alpha) \int_0^{+\infty} \gamma(a)e^{-(\lambda+\mu)a}da \text{ with } \lambda \in \mathbb{C} \text{ and } \operatorname{Re}(\lambda) > -\mu.$

Set
$$\Omega = \{\lambda \in \mathbb{C} : \operatorname{Re}(\lambda) > -\mu\}$$
and consider the map $\Delta : \Omega \to \mathbb{C}$ defined by

(5.5) $\quad\quad\quad \Delta(\lambda) = 1 - \eta(\alpha) \int_0^{+\infty} \gamma(a)e^{-(\lambda+\mu)a}da.$

One can prove that Δ is holomorphic. Moreover, for each $k \geq 1$ and each $\lambda \in \Omega$, we have
$$\frac{d^k \Delta(\lambda)}{d\lambda^k} = (-1)^{k+1} \eta(\alpha) \int_0^{+\infty} a^k \gamma(a) e^{-(\lambda+\mu)a} da.$$

To simplify the notation, we set
$$B_\alpha x = \widehat{A}x + \alpha DH(\overline{v}_\alpha)x \text{ with } D(B_\alpha) = D(\widehat{A})$$
and identify B_α to
$$B_\alpha^{\mathbb{C}}(x + iy) = B_\alpha^{\mathbb{C}} x + iB_\alpha^{\mathbb{C}} y, \quad \forall (x + iy) \in D(B_\alpha^{\mathbb{C}}) := D(\widehat{A}) + iD(\widehat{A}).$$

Note that the part of B_α in $D(B_\alpha)$ is the generator of the linearized equation at $\overline{v}_\alpha$.

LEMMA 5.3. *Let Assumption 5.1 be satisfied. Then the linear operator $B_\alpha : D(\widehat{A}) \subset Y \to Y$ is a Hille-Yosida operator and*
$$\omega_{ess}((B_\alpha)_0) \leq -\mu.$$

PROOF. Since $\alpha DH(\overline{v}_\alpha)$ is a bounded linear operator, it follows that $B_\alpha^{\mathbb{C}}$ is a Hille-Yosida operator. Moreover, by applying Theorem 3 in Thieme [101] (or Theorem 1.2 in [38]) to $B_\alpha + \varepsilon I$ for each $\varepsilon \in (0, \mu)$, we deduce that $\omega_{ess}((B_\alpha)_0) \leq -\mu$. □

LEMMA 5.4. *Let Assumption 5.1 be satisfied. Then the linear operator $B_\alpha : D(\widehat{A}) \subset Y \to Y$ is a Hille-Yosida operator and we have the following:*

(i) $\sigma(B_\alpha^{\mathbb{C}}) \cap \Omega = \{\lambda \in \Omega : \Delta(\lambda) = 0\}.$

(ii) If $\lambda \in \Omega \cap \rho\left(B_\alpha^\mathbb{C}\right)$, we have the following explicit formula for the resolvent

$$\begin{pmatrix} 0 \\ \varphi \end{pmatrix} = \left(\lambda I - B_\alpha^\mathbb{C}\right)^{-1} \begin{pmatrix} \delta \\ \psi \end{pmatrix}$$

$$\Leftrightarrow \varphi(a) = \int_0^a e^{-(\lambda+\mu)(a-s)} \psi(s) ds$$

(5.6)
$$+ \Delta(\lambda)^{-1} \left[\delta + \eta(\alpha) \int_0^{+\infty} \chi_\lambda(s) \psi(s) ds\right] e^{-(\lambda+\mu)a},$$

where

$$\chi_\lambda(s) = \int_s^{+\infty} \gamma(l) e^{-(\lambda+\mu)(l-s)} dl, \quad \forall s \geq 0.$$

PROOF. Assume that $\lambda \in \Omega$ and $\Delta(\lambda) \neq 0$. Then we have

$$\left(\lambda I - B_\alpha^\mathbb{C}\right) \begin{pmatrix} 0 \\ \varphi \end{pmatrix} = \begin{pmatrix} \delta \\ \psi \end{pmatrix}$$

$$\Leftrightarrow \left(\lambda I - \widehat{A}\right) \begin{pmatrix} 0 \\ \varphi \end{pmatrix} = \begin{pmatrix} \delta \\ \psi \end{pmatrix} + \alpha DH\left(\overline{v}_\alpha\right) \begin{pmatrix} 0 \\ \varphi \end{pmatrix}$$

$$\Leftrightarrow \begin{pmatrix} 0 \\ \varphi \end{pmatrix} = \left(\lambda I - \widehat{A}\right)^{-1} \begin{pmatrix} \delta \\ \psi \end{pmatrix} + \left(\lambda I - \widehat{A}\right)^{-1} \alpha DH\left(\overline{v}_\alpha\right) \begin{pmatrix} 0 \\ \varphi \end{pmatrix}$$

$$\Leftrightarrow \varphi(a) = e^{-(\lambda+\mu)a}\delta + \int_0^a e^{-(\lambda+\mu)(a-s)} \psi(s) ds$$

$$+ e^{-(\lambda+\mu)a} \eta(\alpha) \int_0^{+\infty} \gamma(a) \varphi(a) da.$$

Thus

$$\Delta(\lambda) \int_0^{+\infty} \gamma(a) \varphi(a) da = \int_0^{+\infty} \gamma(a) e^{-(\lambda+\mu)a} \delta + \int_0^{+\infty} \gamma(a) \int_0^a e^{-(\lambda+\mu)(a-s)} \psi(s) ds da,$$

so

$$\begin{aligned}\varphi(a) &= e^{-(\lambda+\mu)a} \left[1 + \eta(\alpha) \Delta(\lambda)^{-1} \int_0^{+\infty} \gamma(l) e^{-(\lambda+\mu)l} dl\right] \delta \\ &+ \int_0^a e^{-(\lambda+\mu)(a-s)} \psi(s) ds \\ &+ \eta(\alpha) e^{-(\lambda+\mu)a} \Delta(\lambda)^{-1} \int_0^{+\infty} \gamma(l) \int_0^l e^{-(\lambda+\mu)(l-s)} \psi(s) ds dl.\end{aligned}$$

But we have

$$1 + \eta(\alpha) \Delta(\lambda)^{-1} \int_0^{+\infty} \gamma(a) e^{-(\lambda+\mu)a} = \Delta(\lambda)^{-1}$$

and

$$\int_0^{+\infty} \gamma(l) \int_0^l e^{-(\lambda+\mu)(l-s)} \psi(s) ds dl = \int_0^{+\infty} \int_s^{+\infty} \gamma(l) e^{-(\lambda+\mu)(l-s)} dl \psi(s) ds.$$

Hence (ii) follows. We conclude that

$$\{\lambda \in \Omega : \Delta(\lambda) \neq 0\} \subset \rho\left(\lambda I - B_\alpha^\mathbb{C}\right) \cap \Omega,$$

which implies that
$$\sigma\left(\lambda I - B_\alpha^\mathbb{C}\right) \cap \Omega \subset \{\lambda \in \Omega : \Delta(\lambda) = 0\}.$$
Assume that $\lambda \in \Omega$ is such that $\Delta(\lambda) = 0$. Then for $\varphi(.) = e^{-(\lambda+\mu).}$ we have
$$B_\alpha^\mathbb{C}\begin{pmatrix} 0 \\ \varphi \end{pmatrix} = \lambda \begin{pmatrix} 0 \\ \varphi \end{pmatrix},$$
so $\left(\lambda I - B_\alpha^\mathbb{C}\right)$ is not invertible. We deduce that
$$\{\lambda \in \Omega : \Delta(\lambda) = 0\} \subset \sigma\left(\lambda I - B_\alpha^\mathbb{C}\right) \cap \Omega,$$
and (i) follows. □

The following lemma is well known (see, for example, Dolbeault [**37**, Theorem 2.1.2, p. 43]).

LEMMA 5.5. *Let f be an Holomorphic map from an open connected subset $\Omega \subset \mathbb{C}$ and let $z_0 \in \mathbb{C}$. Then the following assertions are equivalent:*

(i) $f = 0$ on Ω.
(ii) f is null in a neighborhood of z_0.
(iii) For each $k \in \mathbb{N}, f^{(k)}(z_0) = 0$.

LEMMA 5.6. *Let Assumption 5.1 be satisfied. Then we have the following:*

(i) *If $\lambda_0 \in \sigma\left(B_\alpha^\mathbb{C}\right) \cap \Omega$, then λ_0 is isolated in $\sigma\left(B_\alpha^\mathbb{C}\right)$.*

(ii) *If $\lambda_0 \in \sigma\left(B_\alpha^\mathbb{C}\right) \cap \Omega$ and if $k \geq 1$ is the smallest integer such that $\dfrac{d^k \Delta(\lambda_0)}{d\lambda^k} \neq 0$, then λ_0 a pole of order k of $\left(\lambda I - B_\alpha^\mathbb{C}\right)^{-1}$. Moreover, if $k = 1$, then λ_0 is a simple isolated eigenvalue of $B_\alpha^\mathbb{C}$ and the projector on the eigenspace associated to λ_0 is defined by*
$$\widehat{\Pi}_{\lambda_0}\begin{pmatrix} \delta \\ \psi \end{pmatrix} = \begin{pmatrix} 0 \\ \frac{d\Delta(\lambda_0)}{d\lambda}^{-1}\left[\delta + \int_0^{+\infty} \chi_{\lambda_0}(s)\psi(s)ds\right]e^{-(\lambda_0+\mu).} \end{pmatrix}.$$

(iii) *For $\forall x \in Y^\mathbb{C}$,*
$$\overline{\widehat{\Pi}_{\lambda_0} x} = \widehat{\Pi}_{\overline{\lambda_0}} \overline{x}.$$

PROOF. Since Ω is open and connected, we can apply Lemma 5.5 to Δ, and since for each $\lambda > 0$ large enough $\Delta(\lambda) > 0$, we deduce that for each $\lambda \in \Omega$, there exists $m \geq 0$ such that $\dfrac{d^m \Delta(\lambda)}{d\lambda^m} \neq 0$. Moreover, for each $\lambda_0 \in \Omega$, we have
$$\Delta(\lambda) = \sum_{k \geq 0} \frac{(\lambda - \lambda_0)^k}{k!} \frac{d^k \Delta(\lambda_0)}{d\lambda^k}$$
whenever $|\lambda - \lambda_0|$ is small enough. It follows that each root of Δ is isolated. Moreover, assume that there exists $\lambda_0 \in \Omega$ such that $\Delta(\lambda_0) = 0$. Let $m_0 \geq 1$ be the smallest integer such that $\dfrac{d^{m_0}\Delta(\lambda_0)}{d\lambda^{m_0}} \neq 0$. Then we have
$$\Delta(\lambda) = (\lambda - \lambda_0)^{m_0} g(\lambda)$$
with
$$g(\lambda) = \sum_{k=m_0}^{\infty} \frac{(\lambda - \lambda_0)^{k-m_0}}{k!} \frac{d^k \Delta(\lambda_0)}{d\lambda^k}$$

whenever $|\lambda - \lambda_0|$ is small enough. So the multiplicity of λ_0 is k. Now by using Lemma 5.4 we deduce that if $\lambda_0 \in \sigma\left(B_\alpha^\mathbb{C}\right) \cap \Omega$, then λ_0 is isolated in $\sigma\left(B_\alpha^\mathbb{C}\right)$. Moreover, by using (5.6) we deduce that for $k \geq 1$,

$$\lim_{\lambda \to \lambda_0} (\lambda - \lambda_0)^k \left(\lambda I - B_\alpha^\mathbb{C}\right)^{-1} \begin{pmatrix} \delta \\ \psi \end{pmatrix}$$

$$= \lim_{\lambda \to \lambda_0} (\lambda - \lambda_0)^k \Delta(\lambda)^{-1} \left[\delta + \int_0^{+\infty} \chi_\lambda(s)\psi(s)ds\right] \begin{pmatrix} 0 \\ e^{-(\lambda+\mu)\cdot} \end{pmatrix}$$

$$= \lim_{\lambda \to \lambda_0} (\lambda - \lambda_0)^{k-m_0} \frac{1}{g(\lambda)} \left[\delta + \int_0^{+\infty} \chi_\lambda(s)\psi(s)ds\right] \begin{pmatrix} 0 \\ e^{-(\lambda+\mu)\cdot} \end{pmatrix},$$

so

(5.7) $$\lim_{\lambda \to \lambda_0} (\lambda - \lambda_0)^k \left(\lambda I - B_\alpha^\mathbb{C}\right)^{-1} \begin{pmatrix} 0 \\ \psi \end{pmatrix} = 0 \text{ if } k > m_0.$$

But since λ_0 is isolated, we have

$$\left(\lambda I - B_\alpha^\mathbb{C}\right)^{-1} = \sum_{k=-\infty}^{\infty} (\lambda - \lambda_0)^k D_k,$$

where

(5.8) $$D_k = \frac{1}{2\pi i} \int_{S_\mathbb{C}(\lambda_0, \varepsilon)^+} (\lambda - \lambda_0)^{-k-1} \left(\lambda I - B_\alpha^\mathbb{C}\right)^{-1} d\lambda$$

for $\varepsilon > 0$ small enough and each $k \in \mathbb{Z}$. By combining (5.7) and (5.8), we obtain when $\varepsilon \to 0$ that

$$D_{-k} = 0 \text{ for each } k \geq m_0 + 2.$$

It follows that λ_0 is a pole of the resolvent and

$$\left(\lambda I - B_\alpha^\mathbb{C}\right)^{-1} = \sum_{k=-m_0-1}^{\infty} (\lambda - \lambda_0)^k D_k.$$

Noticing that

$$\lim_{\lambda \to \lambda_0} (\lambda - \lambda_0)^{m_0+1} \left(\lambda I - B_\alpha^\mathbb{C}\right)^{-1} = D_{-m_0-1}$$

and using (5.7) once more, we deduce that $D_{-m_0-1} = 0$. Finally, we have

$$\lim_{\lambda \to \lambda_0} (\lambda - \lambda_0)^{m_0} \left(\lambda I - B_\alpha^\mathbb{C}\right)^{-1} = D_{-m_0}$$

and

$$D_{-m_0} \begin{pmatrix} \delta \\ \psi \end{pmatrix} = \frac{1}{g(\lambda_0)} \left[\delta + \int_0^{+\infty} \chi_{\lambda_0}(s)\psi(s)ds\right] \begin{pmatrix} 0 \\ e^{-(\lambda_0+\mu)\cdot} \end{pmatrix}.$$

Therefore, λ_0 is a pole of order $m_0 \geq 1$. □

ASSUMPTION 5.7. Assume that $\alpha^* > 1$ and $\theta^* > 0$ such that $i\theta^*$ and $-i\theta^*$ are simple eigenvalues of B_{α^*} and

$$\sup\{\operatorname{Re}(\lambda) : \lambda \in \sigma(B_{\alpha^*}) \setminus \{i\theta^*, -i\theta^*\}\} < 0.$$

Under Assumption 5.7 we have

$$\overline{\frac{d\Delta(-i\theta^*)}{d\lambda}} = \frac{d\Delta(i\theta^*)}{d\lambda} \neq 0.$$

Moreover, by using assertion (iii) in Lemma 5.6, we can define $\widehat{\Pi}_c : Y \to Y$ as

$$\widehat{\Pi}_c \begin{pmatrix} \delta \\ \varphi \end{pmatrix} = \widehat{\Pi}_{i\theta^*} \begin{pmatrix} \delta \\ \varphi \end{pmatrix} + \widehat{\Pi}_{-i\theta^*} \begin{pmatrix} \delta \\ \varphi \end{pmatrix}, \quad \forall \begin{pmatrix} \delta \\ \varphi \end{pmatrix} \in Y.$$

By using Theorem 3.15 and Lemma 3.2, we deduce the following result.

LEMMA 5.8. *Let Assumptions* 5.1 *and* 5.7 *be satisfied. Then*

$$\sigma \left(B_{\alpha^*} |_{\widehat{\Pi}_c(Y)} \right) = \{i\theta^*, -i\theta^*\}, \sigma \left(B_{\alpha^*} |_{(I-\widehat{\Pi}_c)(Y)} \right) = \sigma(B_{\alpha^*}) \setminus \{i\theta^*, -i\theta^*\},$$

and

$$\omega_0 \left(B_{\alpha^*} |_{(I-\widehat{\Pi}_c)(Y)} \right) < 0.$$

We have

$$\widehat{\Pi}_c \begin{pmatrix} 1 \\ 0 \end{pmatrix} = \begin{bmatrix} 0 \\ \frac{d\Delta(i\theta^*)}{d\lambda}^{-1} e^{-(i\theta^* + \mu)\cdot} + \frac{d\Delta(-i\theta^*)}{d\lambda}^{-1} e^{-(-i\theta^* + \mu)\cdot} \end{bmatrix}$$

$$= \left| \frac{d\Delta(i\theta^*)}{d\lambda} \right|^{-2} \begin{bmatrix} 0 \\ \operatorname{Re}(\Delta(i\theta^*)) \widehat{e}_1 + \operatorname{Im}(\Delta(i\theta^*)) \widehat{e}_2 \end{bmatrix}$$

with

$$\widehat{e}_1 = \left[e^{-(i\theta^* + \mu)\cdot} + e^{-(-i\theta^* + \mu)\cdot} \right], \quad \widehat{e}_2 = \frac{\left(e^{-(i\theta^* + \mu)\cdot} - e^{-(-i\theta^* + \mu)\cdot} \right)}{i}.$$

Set

$$\widehat{\Pi}_s := \left(I - \widehat{\Pi}_c \right).$$

Then we have

$$\widehat{\Pi}_s \begin{pmatrix} 1 \\ 0 \end{pmatrix} = \left(I - \widehat{\Pi}_c \right) \begin{pmatrix} 1 \\ 0 \end{pmatrix}$$

$$= \begin{pmatrix} 1 \\ -\frac{d\Delta(i\theta^*)}{d\lambda}^{-1} e^{-(i\theta^* + \mu)\cdot} - \frac{d\Delta(-i\theta^*)}{d\lambda}^{-1} e^{-(-i\theta^* + \mu)\cdot} \end{pmatrix}$$

$$= \begin{pmatrix} 1 \\ -\left| \frac{d\Delta(i\theta^*)}{d\lambda} \right|^{-2} [\operatorname{Re}(\Delta(i\theta^*)) \widehat{e}_1 + \operatorname{Im}(\Delta(i\theta^*)) \widehat{e}_2] \end{pmatrix}.$$

In order to compute the second derivative of the center manifold at 0, we will need the following lemma.

LEMMA 5.9. *Let Assumptions* 5.1 *and* 5.7 *be satisfied. Then for each* $\lambda \in i\mathbb{R} \setminus \{-i\theta^*, i\theta^*\}$,

$$\left(\lambda I - B_{\alpha^*}^{\mathbb{C}} |_{\widehat{\Pi}_s(Y)} \right)^{-1} \widehat{\Pi}_s \begin{pmatrix} 1 \\ 0 \end{pmatrix}$$

$$= \begin{pmatrix} 0 \\ -\frac{d\Delta(i\theta^*)}{d\lambda}^{-1} \frac{e^{-(i\theta^* + \mu)\cdot}}{(\lambda - i\theta^*)} - \frac{d\Delta(-i\theta^*)}{d\lambda}^{-1} \frac{e^{-(-i\theta^* + \mu)\cdot}}{(\lambda + i\theta^*)} + \Delta(\lambda)^{-1} e^{-(\lambda + \mu)\cdot} \end{pmatrix}.$$

Moreover, if $\lambda = i\theta^*$, *we have*

$$\left(i\theta^* I - B_{\alpha^*}^{\mathbb{C}} |_{\widehat{\Pi}_s(Y)} \right)^{-1} \widehat{\Pi}_s \begin{pmatrix} 1 \\ 0 \end{pmatrix}$$

$$= \begin{pmatrix} 0 \\ -\frac{d\Delta(-i\theta^*)}{d\lambda}^{-1} \frac{e^{-(-i\theta^* + \mu)\cdot}}{2i\theta^*} + \frac{d\Delta(i\theta^*)}{d\lambda}^{-2} \left[\frac{d\Delta(i\theta^*)}{d\lambda} - \frac{1}{2} \frac{d^2\Delta(i\theta^*)}{d\lambda^2} \right] e^{-(i\theta^* + \mu)\cdot} \end{pmatrix}$$

5. HOPF BIFURCATION IN AGE STRUCTURED MODELS

and if $\lambda = -i\theta^*$, we have

$$\left(-i\theta^* I - B_{\alpha^*}^{\mathbb{C}}|_{\widehat{\Pi}_s(Y)}\right)^{-1} \widehat{\Pi}_s \begin{pmatrix} 1 \\ 0 \end{pmatrix}$$

$$= \begin{pmatrix} 0 \\ -\frac{d\Delta(i\theta^*)}{d\lambda}^{-1} \frac{e^{-(i\theta^*+\mu).}}{-2i\theta^*} + \frac{d\Delta(-i\theta^*)}{d\lambda}^{-2}\left[\frac{d\Delta(-i\theta^*)}{d\lambda} - \frac{1}{2}\frac{d^2\Delta(-i\theta^*)}{d\lambda^2}\right]e^{-(-i\theta^*+\mu).} \end{pmatrix}.$$

PROOF. For each $\lambda \in \rho\left(B_{\alpha^*}^{\mathbb{C}}\right)$, we have

$$\left(\lambda I - B_{\alpha^*}^{\mathbb{C}}\right)^{-1} \begin{pmatrix} 0 \\ e^{-(\pm i\theta^*+\mu).} \end{pmatrix} = (\lambda \pm i\theta^*)^{-1} \begin{pmatrix} 0 \\ e^{-(\pm i\theta^*+\mu).} \end{pmatrix}.$$

Hence,

$$\left(\lambda I - B_{\alpha^*}^{\mathbb{C}}|_{\widehat{\Pi}_s(Y)}\right)^{-1} \widehat{\Pi}_s \begin{pmatrix} 1 \\ 0 \end{pmatrix} = \left(\lambda I - B_{\alpha^*}^{\mathbb{C}}\right)^{-1} \widehat{\Pi}_s \begin{pmatrix} 1 \\ 0 \end{pmatrix}$$

$$= \begin{pmatrix} 0 \\ -\frac{d\Delta(i\theta^*)}{d\lambda}^{-1} \frac{e^{-(i\theta^*+\mu).}}{(\lambda - i\theta^*)} - \frac{d\Delta(-i\theta^*)}{d\lambda}^{-1} \frac{e^{-(-i\theta^*+\mu).}}{(\lambda + i\theta^*)} + \Delta(\lambda)^{-1} e^{-(\lambda+\mu).} \end{pmatrix}.$$

Thus,

$$\left(0I - B_{\alpha^*}^{\mathbb{C}}|_{\widehat{\Pi}_s(Y)}\right)^{-1} \widehat{\Pi}_s \begin{pmatrix} 1 \\ 0 \end{pmatrix}$$

$$= \begin{pmatrix} 0 \\ -\frac{d\Delta(i\theta^*)}{d\lambda}^{-1} \frac{e^{-(i\theta^*+\mu).}}{-i\theta^*} - \frac{d\Delta(-i\theta^*)}{d\lambda}^{-1} \frac{e^{-(-i\theta^*+\mu).}}{i\theta^*} + \Delta(0)^{-1} e^{-\mu.} \end{pmatrix}$$

$$= \begin{pmatrix} 0 \\ \left|\frac{d\Delta(i\theta^*)}{d\lambda} i\theta^*\right|^2 \left[\text{Re}\left(\frac{d\Delta(i\theta^*)}{d\lambda} i\theta^*\right) e_1 + \text{Im}\left(\frac{d\Delta(i\theta^*)}{d\lambda} i\theta^*\right) e_2\right] + \Delta(0)^{-1} e^{-\mu.} \end{pmatrix}.$$

Moreover, we have

$$\left(i\theta^* I - B_{\alpha^*}^{\mathbb{C}}|_{\widehat{\Pi}_s(Y)}\right)^{-1} \widehat{\Pi}_s \begin{pmatrix} 1 \\ 0 \end{pmatrix} = \lim_{\substack{\lambda \to i\theta^* \\ \text{with } \lambda \in \rho(B_\alpha^{\mathbb{C}})}} \left(\lambda I - B_{\alpha^*}^{\mathbb{C}}|_{\widehat{\Pi}_s(Y)}\right)^{-1} \widehat{\Pi}_s \begin{pmatrix} 1 \\ 0 \end{pmatrix},$$

so

$$\left(i\theta^* I - B_{\alpha^*}^{\mathbb{C}}|_{\widehat{\Pi}_s(Y)}\right)^{-1} \widehat{\Pi}_s \begin{pmatrix} 1 \\ 0 \end{pmatrix}$$

$$= \lim_{\substack{\lambda \to i\theta^* \\ \text{with } \lambda \in \rho(B_\alpha^{\mathbb{C}})}} \begin{pmatrix} 0 \\ -\frac{d\Delta(i\theta^*)}{d\lambda}^{-1} \frac{e^{-(i\theta^*+\mu).}}{(\lambda - i\theta^*)} - \frac{d\Delta(-i\theta^*)}{d\lambda}^{-1} \frac{e^{-(-i\theta^*+\mu).}}{(\lambda + i\theta^*)} + \Delta(\lambda)^{-1} e^{-(\lambda+\mu).} \end{pmatrix}.$$

Notice that

$$-\frac{d\Delta(i\theta^*)}{d\lambda}^{-1} \frac{e^{-(i\theta^*+\mu).}}{(\lambda - i\theta^*)} + \Delta(\lambda)^{-1} e^{-(\lambda+\mu).}$$

$$= \frac{(\lambda - i\theta^*)^2}{\frac{d\Delta(i\theta^*)}{d\lambda}(\lambda - i\theta^*)\Delta(\lambda)} \frac{\left[-\Delta(\lambda) e^{-(i\theta^*+\mu).} + (\lambda - i\theta^*)\frac{d\Delta(i\theta^*)}{d\lambda} e^{-(\lambda+\mu).}\right]}{(\lambda - i\theta^*)^2}$$

and

$$\frac{(\lambda - i\theta^*)^2}{\frac{d\Delta(i\theta^*)}{d\lambda}(\lambda - i\theta^*)\Delta(\lambda)} = \frac{1}{\frac{d\Delta(i\theta^*)}{d\lambda} \frac{\Delta(\lambda)}{(\lambda - i\theta^*)}} \to \frac{d\Delta(i\theta^*)}{d\lambda}^{-2} \text{ as } \lambda \to i\theta^*.$$

We have
$$\Delta(\lambda) e^{-(i\theta^* + \mu)\cdot} = (\lambda - i\theta^*) \frac{d\Delta(i\theta^*)}{d\lambda} + \frac{(\lambda - i\theta^*)^2}{2} \frac{d^2\Delta(i\theta^*)}{d\lambda^2} + (\lambda - i\theta^*)^3 g(\lambda - i\theta^*)$$

with $g(0) = \frac{1}{3!} \frac{d^2 \Delta(i\theta^*)}{d\lambda^2}$. Therefore,

$$\frac{\left[-\Delta(\lambda) e^{-(i\theta^* + \mu)\cdot} - (\lambda - i\theta^*) \frac{d\Delta(i\theta^*)}{d\lambda} e^{-(\lambda + \mu)\cdot} \right]}{(\lambda - i\theta^*)^2}$$

$$= \frac{-(\lambda - i\theta^*) \frac{d\Delta(i\theta^*)}{d\lambda} \left[e^{-(i\theta^* + \mu)\cdot} - e^{-(\lambda + \mu)\cdot} \right]}{(\lambda - i\theta^*)^2}$$

$$+ \frac{-\left[\frac{(\lambda - i\theta^*)^2}{2} \frac{d^2\Delta(i\theta^*)}{d\lambda^2} + (\lambda - i\theta^*)^3 g(\lambda - i\theta^*) \right] e^{-(i\theta^* + \mu)\cdot}}{(\lambda - i\theta^*)^2}$$

$$\to -\frac{d\Delta(i\theta^*)}{d\lambda} \left(-e^{-(i\theta^* + \mu)\cdot} \right) - \frac{1}{2} \frac{d^2\Delta(i\theta^*)}{d\lambda^2} e^{-(i\theta^* + \mu)\cdot} \text{ as } \lambda \to i\theta^*.$$

Finally, it implies that

$$\left(i\theta^* I - B^{\mathbb{C}}_{\alpha^*} |_{\widehat{\Pi}_s(Y)} \right)^{-1} \widehat{\Pi}_s \begin{pmatrix} 1 \\ 0 \end{pmatrix}$$

$$= \begin{pmatrix} 0 \\ -\frac{d\Delta(-i\theta^*)}{d\lambda}^{-1} \frac{e^{-(-i\theta^* + \mu)\cdot}}{2i\theta^*} + \frac{d\Delta(i\theta^*)}{d\lambda}^{-2} \left[\frac{d\Delta(i\theta^*)}{d\lambda} - \frac{1}{2} \frac{d^2\Delta(i\theta^*)}{d\lambda^2} \right] e^{-(i\theta^* + \mu)\cdot} \end{pmatrix}$$

The case when $\lambda = -i\theta^*$ can be proved similarly. This completes the proof. $\square$

In order to apply the Center Manifold Theorem 4.21 to the above system, we will include the parameter α into the state variable. So we consider the system

$$\begin{cases} \dfrac{dv(t)}{dt} = \widehat{A} v(t) + \alpha(t) H(v(t)), \\ \dfrac{d\alpha(t)}{dt} = 0, \\ v(0) = v_0 \in Y_0, \ \alpha(0) = \alpha_0 \in \mathbb{R}. \end{cases}$$

Making a change of variables

$$\alpha = \widehat{\alpha} + \alpha^* \quad \text{and} \quad v = \widehat{v} + \overline{v}_{\alpha^*},$$

we obtain the system

(5.9) $$\begin{aligned} \frac{d\widehat{v}(t)}{dt} &= \widehat{A}\widehat{v}(t) + (\widehat{\alpha}(t) + \alpha^*) \left[H(\widehat{v}(t) + \overline{v}_{(\widehat{\alpha}(t) + \alpha^*)}) - H(\overline{v}_{(\widehat{\alpha}(t) + \alpha^*)}) \right], \\ \frac{d\widehat{\alpha}(t)}{dt} &= 0. \end{aligned}$$

Set
$$X = Y \times \mathbb{R}, \ X_0 = \overline{D(\widehat{A})} \times \mathbb{R}$$

and
$$\widehat{H}(\widehat{\alpha}, \widehat{v}) = (\widehat{\alpha} + \alpha^*) \left[H(\widehat{v} + \overline{v}_{(\widehat{\alpha} + \alpha^*)}) - H(\overline{v}_{(\widehat{\alpha} + \alpha^*)}) \right].$$

We have
$$\partial_v \widehat{H}(\widehat{\alpha}, \widehat{v})(w) = (\widehat{\alpha} + \alpha^*) DH(\widehat{v} + \overline{v}_{(\widehat{\alpha} + \alpha^*)})(w)$$

and
$$\partial_{\widehat{\alpha}}\widehat{H}(\widehat{\alpha},\widehat{v})(\widetilde{\alpha}) = \widetilde{\alpha}\Big\{ H(\widehat{v}+\overline{v}_{(\widehat{\alpha}+\alpha^*)}) - H(\overline{v}_{(\widehat{\alpha}+\alpha^*)})$$
$$+ (\widehat{\alpha}+\alpha^*)\left[DH(\widehat{v}+\overline{v}_{(\widehat{\alpha}+\alpha^*)})\left(\frac{d\overline{v}_{(\widehat{\alpha}+\alpha^*)}}{d\widehat{\alpha}}\right)\right.$$
$$\left.- DH(\overline{v}_{(\widehat{\alpha}+\alpha^*)})\left(\frac{d\overline{v}_{(\widehat{\alpha}+\alpha^*)}}{d\widehat{\alpha}}\right)\right]\Big\}.$$

So $\partial_v \widehat{H}(0,0) = \alpha^* DH(\overline{v}_{\alpha^*})$ and $\partial_{\widehat{\alpha}} \widehat{H}(0,0) = 0$.

Consider the linear operator $A : D(A) \subset X \to X$ defined by
$$A\begin{pmatrix} \widehat{v} \\ \widehat{\alpha} \end{pmatrix} = \begin{pmatrix} \left(\widehat{A} + \alpha^* DH\left(\overline{v}_{\alpha^*}\right)\right)\widehat{v} \\ 0 \end{pmatrix}$$
with $D(A) = D(\widehat{A}) \times \mathbb{R}$ and the map $F : \overline{D(A)} \to X$ defined by
$$F\begin{pmatrix} v \\ \widehat{\alpha} \end{pmatrix} = \begin{pmatrix} F_1\begin{pmatrix} \widehat{v} \\ \widehat{\alpha} \end{pmatrix} \\ 0_{L^1} \\ 0 \end{pmatrix},$$
where $F_1 : X \to \mathbb{R}$ is defined by
$$F_1\begin{pmatrix} \widehat{v} \\ \widehat{\alpha} \end{pmatrix} = (\widehat{\alpha}+\alpha^*)\left[H(\widehat{v}+\overline{v}_{(\widehat{\alpha}+\alpha^*)}) - H(\overline{v}_{(\widehat{\alpha}+\alpha^*)})\right] - \alpha^* DH\left(\overline{v}_{\alpha^*}\right)(\widehat{v}).$$
Then we have
$$F\begin{pmatrix} 0 \\ \widehat{\alpha} \end{pmatrix} = 0, \ \forall \widehat{\alpha} > 1-\alpha^*, \ \text{ and } DF(0) = 0.$$
Now we can apply Theorem 4.21 to the system
$$(5.10) \qquad \frac{dw(t)}{dt} = Aw(t) + F(w(t)), \quad w(0) = w_0 \in \overline{D(A)}.$$
We have for $\lambda \in \rho(A) \cap \Omega = \Omega \setminus (\sigma(B_{\alpha^*}) \cup \{0\})$ that
$$(\lambda - A)^{-1}\begin{pmatrix} \delta \\ \psi \\ r \end{pmatrix} = \begin{pmatrix} (\lambda - B_{\alpha^*})^{-1}\begin{pmatrix} \delta \\ \psi \end{pmatrix} \\ \frac{r}{\lambda} \end{pmatrix}.$$
By using a similar argument as in the proof of Lemma 5.6 and employing Lemma 5.5, we obtain the following lemma.

LEMMA 5.10. *Let Assumptions 5.1 and 5.7 be satisfied. Then*
$$\sigma(A) = \sigma(B_\alpha) \cup \{0\}.$$
Moreover, the eigenvalues 0 and $\pm i\theta^*$ of A are simple. The corresponding projectors $\Pi_0, \Pi_{\pm i\theta^*} : X + iX \to X + iX$ are defined by
$$\Pi_0 \begin{pmatrix} v \\ r \end{pmatrix} = \begin{pmatrix} 0 \\ r \end{pmatrix},$$
$$\Pi_{\pm i\theta^*}\begin{pmatrix} v \\ r \end{pmatrix} = \begin{pmatrix} \widehat{\Pi}_{\pm i\theta^*} v \\ 0 \end{pmatrix}$$

In this context, the projector $\Pi_c : X \to X$ is defined by
$$\Pi_c(x) = (\Pi_0 + \Pi_{i\theta^*} + \Pi_{-i\theta^*})(x), \quad \forall x \in X.$$
Note that we have
$$\overline{\Pi_{i\theta^*}(x)} = \Pi_{-i\theta^*}(\overline{x}), \quad \forall x \in X + iX,$$
so the above projector Π_c maps X into X. Define the basis of $X_c = R(\Pi_c(X))$ by
$$e_1 = \begin{pmatrix} 0_{\mathbb{R}} \\ e^{-(\mu+i\theta^*)\cdot} + e^{-(\mu-i\theta^*)\cdot} \\ 0_{\mathbb{R}} \end{pmatrix}, \quad e_2 = \begin{pmatrix} 0_{\mathbb{R}} \\ \frac{e^{-(\mu+i\theta^*)\cdot} - e^{-(\mu-i\theta^*)\cdot}}{i} \\ 0_{\mathbb{R}} \end{pmatrix}, \quad e_3 = \begin{pmatrix} 0_{\mathbb{R}} \\ 0_{L^1} \\ 1 \end{pmatrix}$$
and
$$Ae_1 = -\theta^* e_2, \; Ae_2 = \theta^* e_1, \; Ae_3 = 0.$$
Then the matrix of A_c in the basis $\{e_1, e_2, e_3\}$ of X_c is given by

(5.11)
$$M = \begin{bmatrix} 0 & -\theta^* & 0 \\ \theta^* & 0 & 0 \\ 0 & 0 & 0 \end{bmatrix}.$$

Moreover, we have
$$\Pi_c \begin{pmatrix} 1 \\ 0_{L^1} \\ 0_{\mathbb{R}} \end{pmatrix} = \begin{pmatrix} \widehat{\Pi}_{+i\theta^*} \begin{pmatrix} 1 \\ 0_{L^1} \end{pmatrix} + \widehat{\Pi}_{-i\theta^*} \begin{pmatrix} 1 \\ 0_{L^1} \end{pmatrix} \\ 0_{\mathbb{R}} \end{pmatrix}$$
$$= \begin{pmatrix} 0_{\mathbb{R}} \\ \frac{d\Delta(i\theta^*)}{d\lambda}^{-1} e^{-(i\theta^*+\mu)\cdot} + \frac{d\Delta(-i\theta^*)}{d\lambda}^{-1} e^{-(-i\theta^*+\mu)\cdot} \\ 0_{\mathbb{R}} \end{pmatrix}.$$

Thus,
$$\Pi_c \begin{pmatrix} \delta \\ 0_{L^1} \\ r \end{pmatrix} = \delta \left| \frac{d\Delta(i\theta^*)}{d\lambda} \right|^{-2} \left(\operatorname{Re}(\Delta(i\theta^*)) e_1 + \operatorname{Im}(\Delta(i\theta^*)) e_2 \right) + re_3.$$

Therefore, we can apply Theorem 4.21. Let $\Gamma : X_{0c} \to X_{0s}$ be the map defined in Theorem 4.21. Since $X_s \subset Y \times \{0_{\mathbb{R}}\}$ and since $\{e_1, e_2, e_3\}$ is a basis of X_c, it follows that
$$\Psi(x_1 e_1 + x_2 e_2 + x_3 e_3) = \begin{pmatrix} \Psi_1(x_1 e_1 + x_2 e_2 + x_3 e_3) \\ 0_{\mathbb{R}} \end{pmatrix}.$$
Since $F \in C^\infty(X_0, X)$, we can assume that $\Psi \in C_b^3(X_{0c}, X_{0s})$, and the reduced system is given by
$$\frac{dx_c(t)}{dt} = A_0|_{X_c} x_c(t) + \Pi_c F(x_c(t) + \Psi(x_c(t)))$$
$$= A_0|_{X_c} x_c(t) + F_1(x_c(t) + \Psi(x_c(t))) \Pi_c \begin{pmatrix} 1 \\ 0_{L^1} \\ 0_{\mathbb{R}} \end{pmatrix},$$
$$D\Gamma(0) = 0,$$
$$\Gamma \begin{pmatrix} 0_Y \\ \widehat{\alpha} \end{pmatrix} = 0 \text{ for all } \widehat{\alpha} \in \mathbb{R} \text{ with } |\widehat{\alpha}| \text{ small enough.}$$

The system expressed in the basis $\{e_1, e_2, e_3\}$ of X_c is given by

(5.12) $$\frac{d}{dt}\begin{pmatrix} x_1(t) \\ x_2(t) \\ x_3(t) \end{pmatrix} = M \begin{pmatrix} x_1(t) \\ x_2(t) \\ x_3(t) \end{pmatrix} + G\left(x_1(t), x_2(t), x_3(t)\right) V,$$

where M is given by (5.11),

$$V = \left|\frac{d\Delta(i\theta^*)}{d\lambda}\right|^{-2} \begin{pmatrix} Re\left(\Delta(i\theta^*)\right) \\ Im\left(\Delta(i\theta^*)\right) \\ 0 \end{pmatrix}$$

and

$$G(x_1, x_2, x_3) = F_1 \circ (I + \Psi)(x_1 e_1 + x_2 e_2 + x_3 e_3).$$

Here x_3 corresponds to the parameter of the system. Note that we can compute explicitly the third order Taylor expansion of the reduced system around 0. We have

$$DG(x_c) = DF_1(x_c + \Psi(x_c))(I + D\Psi(x_c)),$$
$$D^2 G(x_c)\left(x_c^1, x_c^2\right)$$
$$= D^2 F_1(x_c + \Psi(x_c))\left((I + D\Psi(x_c))\left(x_c^1\right), (I + D\Psi(x_c))\left(x_c^2\right)\right)$$
$$+ DF_1(x_c + \Psi(x_c)) D^2\Psi(x_c)\left(x_c^1, x_c^2\right),$$
$$D^3 G(x_c)\left(x_c^1, x_c^2, x_c^3\right)$$
$$= D^3 F_1(x_c + \Psi(x_c))\left((I + D\Psi(x_c))\left(x_c^1\right), (I + D\Psi(x_c))\left(x_c^2\right), (I + D\Psi(x_c))\left(x_c^3\right)\right)$$
$$+ D^2 F_1(x_c + \Psi(x_c))\left(\left(D^2\Psi(x_c)\right)\left(x_c^1, x_c^3\right), (I + D\Psi(x_c))\left(x_c^2\right)\right)$$
$$+ D^2 F_1(x_c + \Psi(x_c))\left((I + D\Psi(x_c))\left(x_c^1\right), D^2\Psi(x_c)\left(x_c^2, x_c^3\right)\right)$$
$$+ D^2 F_1(x_c + \Psi(x_c))\left(D^2\Psi(x_c)\left(x_c^1, x_c^2\right), (I + D\Psi(x_c))\left(x_c^3\right)\right)$$
$$+ DF_1(x_c + \Psi(x_c)) D^3\Psi(x_c)\left(x_c^1, x_c^2, x_c^3\right).$$

Since $DF_1(0) = 0$, and $\Psi(0) = 0$, $D\Psi(0) = 0$, we obtain

$$DG(0) = 0, \ D^2 G(0)\left(x_c^1, x_c^2\right) = D^2 F_1(0)\left(x_c^1, x_c^2\right)$$

and

$$D^2 G(x_c)\left(x_c^1, x_c^2, x_c^3\right) = D^3 F_1(0)\left(x_c^1, x_c^2, x_c^3\right)$$
$$+ D^2 F_1(0)\left(D^2\Psi(0)\left(x_c^1, x_c^3\right), x_c^2\right)$$
$$+ D^2 F_1(0)\left(x_c^1, D^2\Psi(0)\left(x_c^2, x_c^3\right)\right)$$
$$+ D^2 F_1(0)\left(D^2\Psi(0)\left(x_c^1, x_c^2\right), x_c^3\right).$$

Moreover, by computing the Taylor expansion to the order 3 of the problem, we have

$$G(h) = \frac{1}{2!} D^2 G(0)(h,h) + \frac{1}{3!} D^3 G(0)(h,h,h)$$
$$+ \frac{1}{4!} \int_0^1 (1-t)^4 D^4 F_1(th)(h,h,h,h) dt.$$

Notice that we can compute explicitly that

$$\frac{1}{2!} D^2 G(0)(h,h) + \frac{1}{3!} D^3 G(0)(h,h,h).$$

Because F_1 is explicit, we only need to compute $D^2\Psi(0)$. For each $x, y \in X_c$,

$$D^2\Psi(0)(x,y) = \lim_{\lambda \to +\infty} \int_0^{+\infty} T_{A_0}(l)\Pi_{0s}\lambda(\lambda - A)^{-1} D^{(2)}F(0)\left(e^{-A_{0c}l}x, e^{-A_{0c}l}y\right) dl.$$

Using the fact that

$$\begin{aligned}
e^{A_c t} e_1 &= \cos(\theta^* t) e_1 - \sin(\theta^* t) e_2, \\
e^{A_c t} e_2 &= \sin(\theta^* t) e_1 + \cos(\theta^* t) e_2, \\
e^{A_c t} e_3 &= e_3
\end{aligned}$$

and

$$\cos(\theta^* t) = \frac{\left(e^{i\theta^* t} + e^{-i\theta^* t}\right)}{2}, \quad \sin(\theta^* t) = \frac{\left(e^{i\theta^* t} - e^{-i\theta^* t}\right)}{2i},$$

and following Lemma 5.9 and the same method at the end of Chapter 4 (i.e. the same method as in the proof of (iii) in Theorem 4.21), we can obtain an explicit formula for $D^2\Psi(0)(e_i, e_j)$: For $i, j = 1, 2$,

$$D^2\Psi(0)(e_i, e_j) = \sum_{\substack{\lambda \in \Lambda_{i,j}, \\ k,l=1,2}} \left(c_{ij}(\lambda) \left(\lambda I - B_\alpha^{\mathbb{C}}|_{\widehat{\Pi}_s(Y)}\right)^{-1} \widehat{\Pi}_s \begin{pmatrix} 1 \\ 0_{L_1} \\ 0 \end{pmatrix} D^2 F_1(e_k, e_l) \right),$$

where $\Lambda_{i,j}$ is a finite subset included in $i\mathbb{R}$. So we can compute $D^2\Psi(0)$ and thus have proven that the system (5.12) on the center manifold is C^3 in its variables.

Next, we need to study the eigenvalues of the characteristic equation (5.4). Assume the parameter $\alpha > e$ and consider

$$\Delta(\alpha, \lambda) = 1 - \eta(\alpha) \int_0^{+\infty} \gamma(a) e^{-(\lambda + \mu)a} da$$

with

$$\eta(\alpha) = 1 - \ln(\alpha).$$

We have

$$\frac{\partial \Delta(\alpha, \lambda)}{\partial \alpha} = -\frac{1}{\alpha}\left[\int_0^{+\infty} \gamma(a) e^{-(\lambda+\mu)a} da\right].$$

If $\Delta(\alpha, \lambda) = 0$ and $\alpha > e$, then

$$\frac{\partial \Delta(\alpha, \lambda)}{\partial \alpha} = \frac{1}{\alpha \eta(\alpha)} < 0.$$

In addition to Assumption 5.7, we also make the following assumptions.

ASSUMPTION 5.11. Assume that there is a number $\alpha^* > e$ such that
  a) If $\lambda \in \Omega$ and $\Delta(\alpha, \lambda) = 0$, then $\operatorname{Re}\left(\frac{\partial \Delta(\alpha,\lambda)}{\partial \lambda}\right) > 0$.
  b) There exists a constant $C > 0$ such that for each $\alpha \in [e, \alpha^*]$,
  $$\operatorname{Re}(\lambda) \geq -\mu \text{ and } \Delta(\alpha, \lambda) = 0 \Rightarrow |\lambda| \leq C.$$
  c) There exists $\theta^* > 0$ such that $\Delta(\alpha^*, i\theta^*) = 0$ and $\Delta(\alpha^*, i\theta) \neq 0, \forall \theta \in [0, +\infty) \setminus \{\theta^*\}$.
  d) For each $\alpha \in [e, \alpha^*)$, $\Delta(\alpha, i\theta) \neq 0, \forall \theta \in [0, +\infty)$.

5. HOPF BIFURCATION IN AGE STRUCTURED MODELS

Note that if $\alpha = e$, we have $\Delta(\alpha, \lambda) = 1$, so there is no eigenvalue. By the continuity of $\Delta(\alpha, \lambda)$ and using Assumption 5.11 b), we deduce that there exists $\alpha_1 \in [e, \alpha^*]$ such that

$$\Delta(\alpha, \lambda) \neq 0, \forall \lambda \in \Omega, \forall \alpha \in [e, \alpha_1).$$

Note that because of Assumption 5.11 a), we can apply locally the implicit function theorem and deduce that if $\widehat{\alpha} > e$, $\widehat{\lambda} \in \Omega$, and $\Delta\left(\widehat{\alpha}, \widehat{\lambda}\right) = 0$, then there exist two constants $\varepsilon > 0$, $r > 0$, and a continuously differentiable map $\widehat{\lambda} : (\widehat{\alpha} - \varepsilon, \widehat{\alpha} + \varepsilon) \to \mathbb{C}$, such that

$$\Delta(\alpha, \lambda) = 0 \text{ and } (\alpha, \lambda) \in (\widehat{\alpha} - \varepsilon, \widehat{\alpha} + \varepsilon) \times B_{\mathbb{C}}(0, r) \Leftrightarrow \lambda = \widehat{\lambda}(\alpha).$$

Moreover, we have

$$\Delta\left(\widehat{\alpha}, \widehat{\lambda}(\alpha)\right) = 0$$

and

$$\frac{\partial \Delta\left(\widehat{\alpha}, \widehat{\lambda}(\alpha)\right)}{\partial \alpha} + \frac{\partial \Delta\left(\widehat{\alpha}, \widehat{\lambda}(\alpha)\right)}{\partial \lambda} \frac{d\widehat{\lambda}(\alpha)}{d\alpha} = 0.$$

Thus,

$$\frac{d\widehat{\lambda}(\alpha)}{d\alpha} = \frac{1}{\frac{\partial \Delta(\widehat{\alpha}, \widehat{\lambda}(\alpha))}{\partial \lambda}} \frac{-1}{\alpha \eta(\alpha)}.$$

However,

$$\operatorname{Re}\left(\frac{\partial \Delta\left(\widehat{\alpha}, \widehat{\lambda}(\alpha)\right)}{\partial \lambda}\right) > 0 \Leftrightarrow \operatorname{Re}\left(\frac{1}{\frac{\partial \Delta(\widehat{\alpha}, \widehat{\lambda}(\alpha))}{\partial \lambda}}\right) > 0,$$

so

$$\frac{d\operatorname{Re}\left(\widehat{\lambda}(\alpha)\right)}{d\alpha} > 0.$$

Summarizing the above analysis, we have the following Lemma.

LEMMA 5.12. *Let Assumptions 5.1, 5.7 and 5.11 be satisfied. Then we have the following:*

(a) *For each $\alpha \in [e, \alpha^*)$, the characteristic equation $\Delta(\alpha, \lambda) = 0$ has no roots with positive real part.*

(b) *There exist constants $\varepsilon > 0$, $\eta > 0$, and a continuously differentiable map $\widehat{\lambda} : (\alpha^* - \varepsilon, \alpha^* + \varepsilon) \to \mathbb{C}$, such that*

$$\Delta\left(\alpha, \widehat{\lambda}(\alpha)\right) = 0, \quad \forall \alpha \in (\alpha^* - \varepsilon, \alpha^* + \varepsilon)$$

with

$$\widehat{\lambda}(\alpha^*) = i\theta^* \text{ and } \frac{d}{d\alpha}\operatorname{Re}\left(\widehat{\lambda}(\alpha^*)\right) > 0,$$

and for each $\alpha \in (\alpha^* - \varepsilon, \alpha^* + \varepsilon)$, if

$$\Delta(\alpha, \lambda) = 0, \lambda \neq \widehat{\lambda}(\alpha), \text{ and } \lambda \neq \overline{\widehat{\lambda}(\alpha)},$$

then

$$\operatorname{Re}(\lambda) < -\eta.$$

In order to find the critical values of the parameter α and verify the transversality condition, we need to be more specific about the function $\gamma(a)$. We make the following assumption.

ASSUMPTION 5.13. Assume that

(5.13) $$\gamma(a) = \begin{cases} \delta(a-\tau)^n e^{-\zeta(a-\tau)}, & \text{if } a \geq \tau \\ 0, & \text{if } a \in [0, \tau) \end{cases}$$

for some integer $n \geq 1$, $\tau \geq 0$, $\zeta > 0$, and

$$\delta = \left(\int_\tau^{+\infty} (a-\tau)^n e^{-\zeta(a-\tau)} da \right)^{-1} > 0.$$

Note that if $n \geq 1$, then γ satisfies the conditions in Assumption 5.1. We have for $\lambda \in \Omega$ that

$$\begin{aligned} \int_0^{+\infty} \gamma(a) e^{-(\mu+\lambda)a} da &= \int_\tau^{+\infty} \gamma(a) e^{-(\mu+\lambda)a} da \\ &= \delta e^{-(\mu+\lambda)\tau} \int_\tau^{+\infty} (a-\tau)^n e^{-(\mu+\zeta+\lambda)(a-\tau)} da \\ &= \delta e^{-(\mu+\lambda)\tau} \int_0^{+\infty} l^n e^{-(\mu+\zeta+\lambda)l} dl. \end{aligned}$$

Set

$$I_n(\lambda) = \int_0^{+\infty} l^n e^{-(\mu+\zeta+\lambda)l} dl \quad \text{for each } n \geq 0 \text{ and each } \lambda \in \Omega.$$

Then we have

$$\begin{aligned} \Delta(\alpha, \lambda) &= 1 - \eta(\alpha) \int_0^{+\infty} \gamma(a) e^{-(\lambda+\mu)a} da \\ &= 1 - \eta(\alpha) \delta e^{-(\mu+\lambda)\tau} I_n(\lambda). \end{aligned}$$

Then by integrating by part we have for $n \geq 1$ that

$$\begin{aligned} I_n(\lambda) &= \int_0^{+\infty} l^n e^{-(\mu+\zeta+\lambda)l} dl \\ &= \left[\frac{l^n e^{-(\mu+\zeta+\lambda)l}}{-(\mu+\zeta+\lambda)} \right]_0^{+\infty} - \int_0^{+\infty} \frac{n l^{n-1} e^{-(\mu+\zeta+\lambda)l}}{(\mu+\zeta+\lambda)} dl \\ &= \frac{n}{(\mu+\zeta+\lambda)} I_{n-1}(\lambda) \end{aligned}$$

and

$$I_0(\lambda) = \int_0^{+\infty} e^{-(\mu+\zeta+\lambda)l} dl = \frac{1}{(\mu+\zeta+\lambda)}.$$

Therefore,

$$I_n(\lambda) = \frac{n!}{(\mu+\zeta+\lambda)^{n+1}}, \quad \forall n \geq 0$$

with $0! = 1$.

The characteristic equation (5.4) becomes

(5.14) $$1 = \eta(\alpha) \delta n! \frac{e^{-\tau(\mu+\zeta+\lambda)}}{(\mu+\zeta+\lambda)^{n+1}}, \quad \text{Re}(\lambda) > -\mu.$$

5. HOPF BIFURCATION IN AGE STRUCTURED MODELS

Note that when $n = 0$, the above characteristic equation (5.14) is well known in the context of delay differential equation (see Hale and Verduyn Lunel [51], p.341). Note also that when $\tau = 0$, (5.14) becomes trivial. Indeed, assume that $\tau = 0$ and $\eta < 0$, then we have

$$(\mu + \zeta + \lambda)^{n+1} = -|\eta|\, \delta n! = |\eta|\, \delta n! e^{i(2k+1)\pi} \text{ for } k = 0, 1, 2, \ldots$$

so

$$\lambda = -(\mu + \zeta) + \sqrt[n+1]{|\eta|\, \delta n!}\, e^{i\frac{(2k+1)}{n+1}\pi} \text{ for } k = 0, 1, 2, \ldots$$

Note that

$$\begin{aligned}
\frac{d\Delta(\lambda)}{d\lambda} &= \eta \int_0^{+\infty} a\gamma(a) e^{-(\lambda+\mu)a} da \\
&= \eta \delta e^{-(\lambda+\mu)\tau} \int_\tau^{+\infty} a(a-\tau)^n e^{-(\mu+\zeta+\lambda)(a-\tau)} da \\
&= \eta \delta e^{-(\lambda+\mu)\tau} \left[\int_\tau^{+\infty} (a-\tau)^{n+1} e^{-(\mu+\zeta+\lambda)(a-\tau)} da \right. \\
&\qquad \left. + \tau \int_\tau^{+\infty} (a-\tau)^n e^{-(\mu+\zeta+\lambda)(a-\tau)} da \right] \\
&= \eta \delta e^{-(\lambda+\mu)\tau} [I_{n+1} + \tau I_n] \\
&= \eta \delta e^{-(\lambda+\mu)\tau} \left[\frac{n+1}{(\mu+\zeta+\lambda)} + \tau \right] I_n \\
&= \left[\frac{n+1}{(\mu+\zeta+\lambda)} + \tau \right] [1 - \Delta(\lambda)].
\end{aligned}$$

If $\Delta(\lambda) = 0$, it follows that

$$\frac{d\Delta(\lambda)}{d\lambda} = \left[\frac{n+1}{(\mu+\zeta+\lambda)} + \tau \right] \neq 0 \text{ and } \operatorname{Re}\left(\frac{d\Delta(\lambda)}{d\lambda}\right) > 0.$$

Hence, all eigenvalues are simple and we can apply the implicit function theorem around each solution of the characteristic equation.

Note that

$$|\mu + \zeta + \lambda|^2 = |\eta(\alpha)\, \delta n!|^{\frac{2}{n+1}} e^{-\frac{2\tau}{n+1}(\mu+\zeta+\operatorname{Re}(\lambda))}.$$

So

(5.15) $$\operatorname{Im}(\lambda)^2 = |\eta(\alpha)\, \delta n!|^{\frac{2}{n+1}} e^{-\frac{2\tau}{n+1}(\mu+\zeta+\operatorname{Re}(\lambda))} - (\mu + \zeta + \operatorname{Re}(\lambda))^2.$$

Thus, there exists $\delta_1 > 0$ such that $-\mu < \operatorname{Re}(\lambda) \leq \delta_1$. This implies that the characteristic equation (5.14) satisfies Assumption 5.11 b). Using (5.15) we also know that for each real number δ, there is at most one pair of complex conjugate eigenvalues such that $\operatorname{Re}(\lambda) = \delta$.

LEMMA 5.14. *Let Assumption* 5.13 *be satisfied. Then Assumptions* 5.1, 5.7 *and* 5.11 *are satisfied.*

PROOF. In order to prove the above lemma it is sufficient to prove that for $\alpha > e$ large enough there exists $\lambda \in \mathbb{C}$ such that

$$\Delta(\alpha, \lambda) = 0 \text{ and } \operatorname{Re}(\lambda) > 0.$$

The characteristic equation can be rewritten as follows

$$(\xi + \lambda)^{n+1} = -\chi(\alpha) e^{-\tau(\xi+\lambda)}, \ \operatorname{Re}(\lambda) \geq 0,$$

where
$$\chi(\alpha) = (\ln(\alpha) - 1)\,\delta n! = \ln\left(\frac{\alpha}{e}\right)\delta n! > 0 \text{ and } \xi = \mu + \zeta > 0.$$

Replacing λ by $\widehat{\lambda} = \tau(\xi + \lambda)$ and $\chi(\alpha)$ by $\widehat{\chi}(\alpha) = \tau^{n+1}\chi(\alpha)$, we obtain
$$\widehat{\lambda}^{n+1} = -\widehat{\chi}(\alpha)\,e^{-\widehat{\lambda}} \text{ and } \operatorname{Re}\left(\widehat{\lambda}\right) \geq \tau\xi.$$
$$\Leftrightarrow \quad \widehat{\lambda}^{n+1} = \widehat{\chi}(\alpha)\,e^{-\widehat{\lambda}+(2k+1)\pi i} \text{ and } \operatorname{Re}\left(\widehat{\lambda}\right) \geq \tau\xi, k \in \mathbb{Z}.$$

So we must find $\widehat{\lambda} = a + ib$ with $a > \tau\xi$ such that
$$\begin{cases} a = \widehat{\chi}(\alpha)^{\frac{1}{n+1}}\,e^{-a}\cos\left(\frac{b+(2k+1)\pi}{n+1}\right), \\ b = \widehat{\chi}(\alpha)^{\frac{1}{n+1}}\,e^{-a}\sin\left(-\frac{b+(2k+1)\pi}{n+1}\right) \end{cases}$$
for some $k \in \mathbb{Z}$.

From the first equation of the above system we must have
$$\frac{a}{\widehat{\chi}(\alpha)^{\frac{1}{n+1}}\,e^{-a}} \in [0,1) \text{ and } \cos\left(\frac{b+(2k+1)\pi}{n+1}\right) > 0.$$

Moreover, the above system can also be written as
$$\tan\left(\frac{b+(2k+1)\pi}{n+1}\right) = -\frac{b}{a},$$
and
$$ae^{a} = \widehat{\chi}(\alpha)^{\frac{1}{n+1}}\cos\left(\frac{b+(2k+1)\pi}{n+1}\right).$$

We set
$$\widehat{b} = \frac{b+(2k+1)\pi}{n+1}.$$
Then
$$b = (n+1)\widehat{b} - (2k+1)\pi.$$
The problem becomes to find $\widehat{\theta} \in \mathbb{R}\setminus\left\{\frac{\pi}{2} + m\pi : m \in \mathbb{Z}\right\}$ such that

(5.16) $\qquad \cos(\widehat{\theta}) > 0, \quad \tan\left(\widehat{\theta}\right) = -\dfrac{(n+1)\widehat{\theta} - (2k+1)\pi}{a}, \quad k \in \mathbb{Z},$

and

(5.17) $\qquad ae^{a} = \widehat{\chi}(\alpha)^{\frac{1}{n+1}}\cos\left(\widehat{\theta}\right).$

Fix $a > \tau\xi = \tau(\mu+\xi)$, then it is clear that we can find $\widehat{\theta} \in [-\frac{\pi}{2}, \frac{\pi}{2}]$ such that (5.16) is satisfied. Moreover, $\widehat{\chi}(e) = 0$ and $\widehat{\chi}(\alpha) \to +\infty$ as $\alpha \to +\infty$. Thus, we can find $\widehat{\alpha} > e$, in turn we can $\alpha > e$, such that (5.17) is satisfied. The result follows. □

Therefore, by the Hopf bifurcation theorem (see Hassard et al. [52]) and Proposition 4.22 we have the following result.

PROPOSITION 5.15. *Let Assumptions 5.1 and 5.13 be satisfied. Then there exists $\alpha^* > 0$, where α^* satisfies Assumption 5.7, such that the age structured model (5.1) undergoes a Hopf bifurcation at the equilibrium $v = \bar{v}_{\alpha^*}$ given by (5.3). In particular, a non-trivial periodic solution bifurcates from the equilibrium $v = \bar{v}_{\alpha^*}$ when $\alpha = \alpha^*$.*

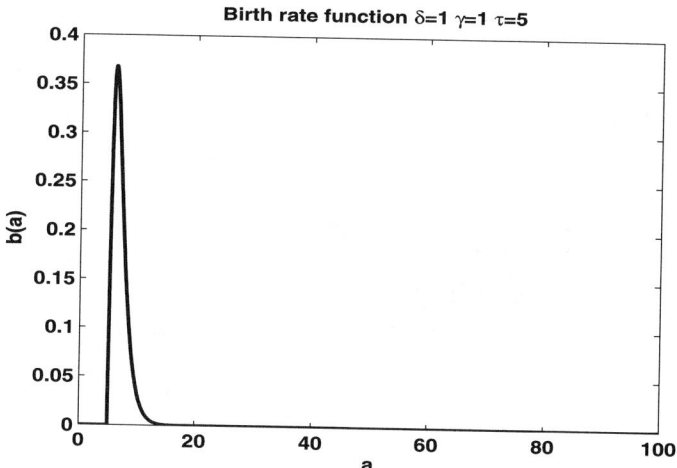

FIGURE 5.1. The birth rate function $b(a)$ with $\delta = 1, \gamma = 1$, and $\tau = 5$.

To carry out some numerical simulations, we consider the equation
$$\begin{cases} \dfrac{\partial u}{\partial t} + \dfrac{\partial u}{\partial a} = -\mu u(t,a), \ t \geq 0, \ a \geq 0 \\ u(t,0) = h\left(\int_0^{+\infty} b(a)u(t,a)da\right) \\ u(0,a) = u_0(a) \end{cases}$$

with the initial value function
$$u_0(a) = a\exp(-0.08a),$$
the fertility rate function
$$h(x) = \alpha x \exp(-\beta x)$$
and the birth rate function (see Figure 5.1)
$$b(a) = \begin{cases} \delta \exp(-\gamma(a-\tau))(a-\tau), & \text{if } a \geq \tau, \\ 0, & \text{if } a \in [0,\tau]. \end{cases}$$
where
$$\mu = 0.1, \ \beta = 1, \ \delta = 1, \ \gamma = 1, \ \tau = 5.$$
The equilibrium is given by
$$\bar{u}(a) = Ce^{-\mu a}, \ a \geq 0, \ C = h\left(\int_0^{+\infty} b(a)e^{-\mu a}Cda\right).$$

We choose $\alpha \geq 0$ as the bifurcation parameter. When $\alpha = 10$, the solution converges to the equilibrium (see Figure 5.2 upper figure). When $\alpha = 20$, the equilibrium loses its stability, a Hopf bifurcation occurs and there is a time periodic solution (see Figure 5.2 lower figure).

Age structured models have been used to study many biological and epidemiological problems, such as the evolutionary epidemiology of type A influenza (Pease [86], Castillo-Chavez et al. [13], Inaba [60, 62]), the epidemics of schistosomiasis in human hosts (Zhang et al. [114]), population dynamics (Gurtin and MacCamy

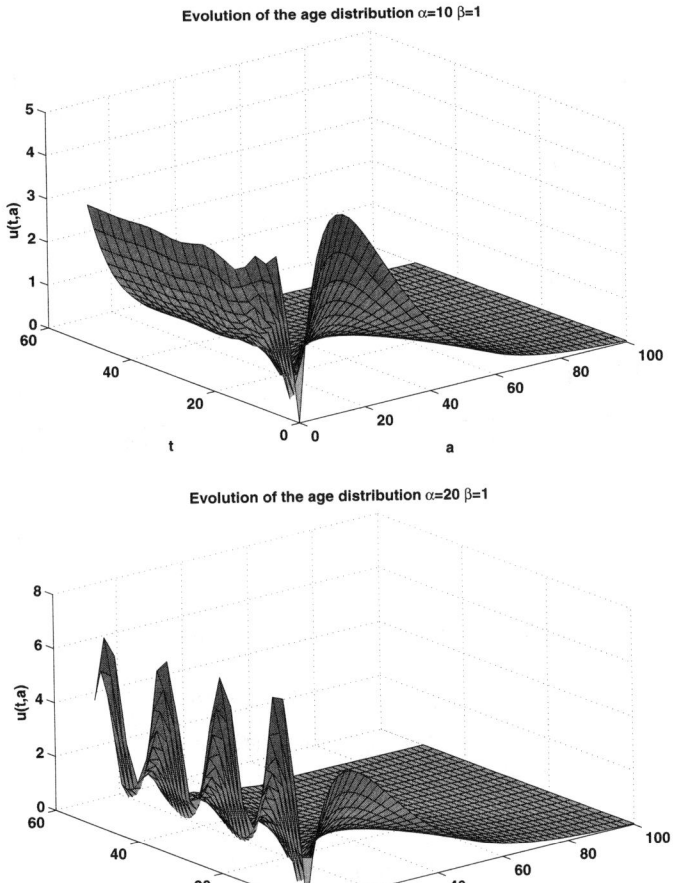

FIGURE 5.2. The age distribution of $u(t,a)$, which converges to the equilibrium when $\alpha = 10$ (upper) and is time periodic when $\alpha = 20$ (lower).

[**46**], Webb [**107, 108**], Iannelli [**59**], Cushing [**27**]), and the epidemics of antibiotic-resistant bacteria in hospitals (D'Agata et al. [**29, 28**], Webb et al. [**109**]). Periodic solutions have been observed in some of these age structured models (Castillo-Chavez et al. [**13**], Inaba [**60, 62**], Zhang et al. [**114**]) and it is believed that such periodic solutions are induced by Hopf bifurcation (Cushing [**25, 26**], Prüss [**89**], Swart [**96**], Kostava and Li [**67**], Bertoni [**10**]). In this chapter, we established a Hopf bifurcation theorem for the age structured model (5.1). Recently, we (Magal and Ruan [**79**]) also studied Hopf bifurcation in an evolutionary epidemiological model of type A influenza (Pease [**86**] and Inaba [**60, 62**]). We think that the center manifold theorem (Theorem 4.21) and the techniques used in analyzing (5.1) can be developed to investigate Hopf bifurcations in some of the above mentioned biological and epidemiological models with age structure (for example, the schistosomiasis model in Zhang et al. [**114**]) and some other structured models (Hoppensteadt

[**57**], Webb [**108**], Iannelli [**59**], Cushing [**27**], Magal and Ruan [**77**]). It may also be employed to study the stability change in age structured SIR epidemic models (Thieme [**100**], Andreasen [**2**], Cha et al. [**14**]).

Acknowledgments

The authors are grateful to the referee for his helpful comments and suggestions.

Bibliography

[1] R. M. Anderson, Discussion: The Kermack-McKendrick epidemic threshold theorem, *Bull. Math. Biol.* **53** (1991), 3-32.

[2] V. Andreasen, Instability in an SIR-model with age-dependent susceptibility, in *"Mathematical Population Dynamics"*, Vol. **1**, eds. by O. Arino, D. Axelrod, M. Kimmel and M. Langlais, Wuerz Publishing, Winnipeg, 1995, pp. 3-14.

[3] W. Arendt, Resolvent positive operators, *Proc. London Math. Soc.* **54** (1987), 321-349. MR872810 (88c:47074)

[4] W. Arendt, Vector valued Laplace transforms and Cauchy problems, *Israel J. Math.* **59** (1987), 327-352. MR920499 (89a:47064)

[5] W. Arendt, C. J. K. Batty, M. Hieber, and F. Neubrander, *Vector-Valued Laplace Transforms and Cauchy Problems*, Birkhäuser, Basel, 2001. MR1886588 (2003g:47072)

[6] A. Avez, *Calcul différentiel*, Masson, Paris, 1983. MR700398 (85c:58001)

[7] J. M. Ball, Saddle point analysis for an ordinary differential equation in a Banach space and an application to dynamic buckling of a beam, in *"Nonlinear Elasticity"*, ed. by R. W. Dickey, Academic Press, New York, 1973, pp. 93-160.

[8] P. W. Bates and C. K. R. T. Jones, Invariant manifolds for semilinear partial differential equations, *Dynamics Reported*, ed. by U. Kirchgraber and H. O. Walther, Vol. **2**, John Wiley & Sons, 1989, pp. 1-38. MR1000974 (90g:58017)

[9] P. W. Bates, K. Lu and C. Zeng, Existence and persistence of invariant manifolds for semi flows in Banach space, *Mem. Amer. Math. Soc.* **135** (1998), No. 645. MR1445489 (99b:58210)

[10] S. Bertoni, Periodic solutions for non-linear equations of structure populations, *J. Math. Anal. Appl.* **220** (1998), 250-267. MR1613952 (98m:35209)

[11] F. E. Browder, On the spectral theory of elliptic differential operators, *Math. Ann.* **142** (1961), 22-130. MR0209909 (35:804)

[12] J. Carr, *Applications of Centre Manifold Theory*, Springer-Verlag, New York, 1981. MR635782 (83g:34039)

[13] C. Castillo-Chavez, H. W. Hethcote, V. Andreasen, S. A. Levin, and W. M. Liu, Epidemiological models with age structure, proportionate mixing, and cross-immunity, *J. Math. Biol.* **27** (1989), 233-258. MR1000090 (90g:92056)

[14] Y. Cha, M. Iannelli and F. A. Milner, Stability change of an epidemic model, *Dynam. Syst. Appl.* **9** (2000), 361-376. MR1844637 (2002f:92028)

[15] C. Chicone and Y. Latushkin, Center manifolds for infinite dimensional nonautonomous differential equations, *J. Differential Equations* **141** (1997), 356-399. MR1488358 (2000i:34117)

[16] S.-N. Chow and J. K. Hale, *Methods of Bifurcation Theory*, Springer-Verlag, New York, 1982. MR660633 (84e:58019)

[17] C.-N. Chow, C. Li and D. Wang, *Normal Forms and Bifurcation of Planar Vector Fields*, Cambridge Univ. Press, Cambridge, 1994. MR1290117 (95i:58161)

[18] C.-N. Chow, X.-B. Lin and K. Lu, Smooth invariant foliations in infinite dimensional spaces, *J. Differential Equations* **94** (1991), 266-291. MR1137616 (92k:58210)

[19] S.-N. Chow, W. Liu, and Y. Yi, Center manifolds for smooth invariant manifolds, *Trans. Amer. Math. Soc.* **352** (2000), 5179-5211. MR1650077 (2001b:37032)

[20] S.-N. Chow, W. Liu and Y. Yi, Center manifolds for invariant sets, *J. Differential Equations* **168** (2000), 355-385. MR1808454 (2002a:37028)

[21] S.-N. Chow and K. Lu, Invariant manifolds for flows in Banach spaces, *J. Differential Equations* **74** (1988), 285-317. MR952900 (89h:58163)

[22] S.-N. Chow and K. Lu, C^k centre unstable manifolds, *Proc. Royal Soc. Edinburgh* **108A** (1988), 303-320. MR943805 (90a:58148)

[23] S.-N. Chow and K. Lu, Invariant manifolds and foliations for quasiperiodic systems, *J. Differential Equations* **117** (1995), 1-27. MR1320181 (96b:34064)

[24] S. N. Chow and Y. Yi, Center manifold and stability for skew-product flows, *J. Dynam. Differential Equations* **6** (1994), 543-582. MR1303274 (95k:58142)

[25] J. M. Cushing, Model stability and instability in age structured populations, *J. Theoret. Biol.* **86** (1980), 709-730. MR596373 (81m:92039)

[26] J. M. Cushing, Bifurcation of time periodic solutions of the McKendrick equations with applications to population dynamics, *Comput. Math. Appl.* **9** (1983), 459-478. MR702665 (84g:92028)

[27] J. M. Cushing, *An Introduction to Structured Population Dynamics*, SIAM, Philadelphia, 1998. MR1636703 (99k:92024)

[28] E. M. C. D'Agata, P. Magal, D. Olivier, S. Ruan and G. F. Webb, Modeling antibiotic resistance in hospitals: The impact of minimizing treatment duration, *J. Theoretical Biology* **249** (2007), 487-499.

[29] E. M. C. D'Agata, P. Magal, S. Ruan and G. F. Webb, Asymptotic behavior in nosocomial epidemi models with antibiotic resistance, *Differential Integral Equations* **19** (2006), 573-600. MR2235142 (2008a:35153)

[30] G. Da Prato and A. Lunardi, Stability, instability and center manifold theorem for fully nonlinear autonomous parabolic equations in Banach spaces, *Arch. Rational Mech. Anal.* **101** (1988), 115-141. MR921935 (89e:35019)

[31] G. Da Prato and E. Sinestrari, Differential operators with non-dense domain, *Ann. Scuola. Norm. Sup. Pisa Cl. Sci.* **14** (1987), 285-344. MR939631 (89f:47062)

[32] O. Diekmann, H. Heesterbeek and H. Metz, The legacy of Kermack and McKendrick, in *"Epidemic Models: Their Strcture and Relation to Data"*, ed. by. D. Mollison, Cambridge University Press, Cambridge, 1995, pp. 95-115.

[33] O. Diekmann, P. Getto and M. Gyllenberg, Stability and bifurcation analysis of Volterra functional equations in the light of suns and stars, *SIAM J. Math. Anal.* **34** (2007), 1023-1069. MR2368893 (2008j:45002)

[34] O. Diekmann and S. A. van Gils, Invariant manifold for Volterra integral equations of convolution type, *J. Differential Equations* **54** (1984), 139-180. MR757290 (85h:45026)

[35] O. Diekmann and S. A. van Gils, The center manifold for delay equations in the light of suns and stars, in *"Singularity Theory and its Applications,"*, Lect. Notes Math. Vol. **1463**, Springer, Berlin, 1991, pp. 122-141. MR1129051 (92m:47151)

[36] O. Diekmann, S. A. van Gils, S. M. Verduyn Lunel, and H.-O. Walther, *Delay Equations. Functional-, Complex-, and Nonlinear Analysis,* Springer-Verlag, New York, 1995. MR1345150 (97a:34001)

[37] P. Dolbeault, *Analyse Complexe*, Masson, Paris, 1990. MR1059456 (91h:30001)

[38] A. Ducrot, Z. Liu and P. Magal, Essential growth rate for bounded linear perturbation of non densely defined Cauchy problems, *J. Math. Anal. Appl.* **341** (2008), 501-518. MR2394101 (2008m:34130)

[39] A. Ducrot, Z. Liu and P. Magal, Projectors on the generalized eigenspaces for neutral functional differential equations in L^p spaces, *Can. J. Math.* (to appear).

[40] N. Dunford and J. T. Schwartz, *Linear Operator, Part I: General Theory*, Interscience, New York, 1958. MR1009162 (90g:47001a)

[41] K.-J. Engel and R. Nagel, *One Parameter Semigroups for Linear Evolution Equations*, Springer-Verlag, New York, 2000. MR1721989 (2000i:47075)

[42] K. Ezzinbi and M. Adimy, The basic theory of abstract semilinear functional differential equations with non-dense domain, in *"Delay Differential Equations and Applications"*, eds. by O. Arino, M. L. Hbid and E. Ait Dads, Springer, Berlin, 2006, pp. 347-397. MR2337821

[43] T. Faria, W. Huang and J. Wu, Smoothness of center manifolds for maps and formal adjoints for semilinear FDES in general Banach spaces, *SIAM J. Math. Anal.* **34** (2002), 173-203. MR1950831 (2003k:34146)

[44] N. Fenichel, Geometric singular perturbation theory for ordinary differential equations, *J. Differential Equations* **31** (1979), 53-98. MR524817 (80m:58032)

[45] Th. Gallay, A center-stable manifold theorem for differential equations in Banach spaces, *Comm. Math. Phys.* **152** (1993), 249-268. MR1210168 (94i:34126)

[46] M. E. Gurtin and R. C. MacCamy, Nonlinear age-dependent population dynamics, *Arch. Rational Mech. Anal.* **54** (1974), 28l-300. MR0354068 (50:6550)

[47] J. Hadamard, Sur l'iteration et les solutions asymptotiques des equations differentielles, *Bull. Soc. Math. France* **29** (1901), 224-228.

[48] J. K. Hale, Integral manifolds of perturbated differential equations, *Ann. Math.* **73** (1961), 496-531. MR0123786 (23:A1108)

[49] J. K. Hale, *Ordinary Differential Equations*, 2nd Ed., Krieger Pub., Huntington, NY, 1980. MR587488 (82e:34001)

[50] J. K. Hale, Flows on center manifolds for scalar functional differential equations, *Proc. Roy. Soc. Edinburgh* **101A** (1985), 193-201. MR824220 (87d:34117)

[51] J. K. Hale and S. M. Verduyn Lunel, *Introduction to Functional Differential Equations*, Springer-Verlag, New York, 1993. MR1243878 (94m:34169)

[52] B. D. Hassard, N. D. Kazarinoff and Y.-H. Wan, *Theory and Applications of Hopf Bifurcaton*, Cambridge Univ. Press, Cambridge, 1981. MR603442 (82j:58089)

[53] K. P. Hadeler and K. Dietz, Nonlinear hyperbolic partial differential equations for the dynamics of parasite populations, *Comput. Math. Appl.* **9** (1983), 415-430. MR702661 (84h:92031)

[54] D. Henry, *Geometric Theory of Semilinear Parabolic Equations*, Lect. Notes Math. Vol. **840**, Springer-Verlag, Berlin, 1981. MR610244 (83j:35084)

[55] M. Hirsch, C. Pugh and M. Shub, *Invariant Manifolds*, Lecture Notes in Math. Vol. **583**, Springer-Verlag, New York, 1976. MR0501173 (58:18595)

[56] A. Homburg, Global aspects of homoclinic bifurcations of vector fields, *Mem. Amer. Math. Soc.* **121** (1996), No. 578. MR1327210 (96i:58125)

[57] F. Hoppensteadt, *Mathematical Theories of Populations: Demographics, Genetics, and Epidemics*, SIAM, Philadelphia, 1975. MR0526771 (58:26164)

[58] H. J. Hupkes and S. M. Verduyn Lunel, Center manifold theory for functional differential equations of mixed type, *J. Dynam. Differential Equations* **19** (2007), 497-560. MR2333418 (2008c:34156)

[59] M. Iannelli, *Mathematical Theory of Age-Structured Population Dynamics*, Appl. Math. Monographs C. N. R., Vol. **7**, Giadini Editori e Stampatori, Pisa, 1994.

[60] H. Inaba, Mathematical analysis for an evolutionary epidemic model, in *"Mathematical Models in Medical and Health Sciences"*, eds. by M. A. Horn, G. Simonett and G. F. Webb, Vanderbilt Univ. Press, Nashville, TN, 1998, pp. 213-236. MR1741583 (2001g:92037)

[61] H. Inaba, Kermack and McKendrick revisted: The variable susceptibility model for infectious diseases, *Japan J. Indust. Appl. Math.* **18** (2001), 273-292. MR1842912 (2002e:92025)

[62] H. Inaba, Endemic threshold and stability in an evolutionary epidemic model, in *"Mathematical Approaches for Emerging and Reemerging Infectious Diseases: Models, Methods, and Theory"*, eds. by C. Castillo-Chavez et al., Springer-Verlag, New York, 2002, pp. 337-359. MR1938912

[63] T. Kato, *Perturbation Theory for Linear Operators*, Springer-Verlag, Berlin, 1995. MR1335452 (96a:47025)

[64] H. Kellermann and M. Hieber, Integrated semigroups, *J. Funct. Anal.* **84** (1989), 160-180. MR999494 (90h:47072)

[65] A. Kelley, The stable, center-stable, center, center-unstable, unstable manifolds. *J. Differential Equations* **3** (1967), 546-570. MR0221044 (36:4096)

[66] B. L. Keyfitz and N. Keyfitz, The McKendrick partial differential equation and its uses in epidemiology and population study, *Math. Comput. Modelling* **26** (1997), No. 6, l-9. MR1601714

[67] T. Kostava and J. Li, Oscillations and stability due to juvenile competitive effects on adult fertility, *Comput. Math. Appl.* **32** (1996), No. 11, 57-70.

[68] T. Krisztin, Invariance and noninvarince of center manifolds of time-t maps with respect to the semiflow, *SIAM J. Math. Anal.* **36** (2004), 717-739. MR2111913 (2005h:34211)

[69] N. Krylov and N. N. Bogoliubov, *The Application of Methods of Nonlinear Mechanics to the Theory of Stationary Oscillations*, Pub. **8** Ukrainian Acad. Sci., Kiev, 1934.

[70] S. Lang, *Real Analysis*, 2nd Ed., Addison-Wesley, Reading, MA, 1983. MR783635 (87b:00001)

[71] A. M. Liapunov, Probléme génerale de la stabilité du mouvement, *Ann. Fac. Sci. Toulouse* **2** (1907), 203-474. MR0021186 (9:34j)

[72] X.-B. Lin, Homoclinic bifurcations with weakly expanding center manifolds, *Dynamics Reported* (new series), ed. by C. K. R. T. Jones, U. Kirchgraber and H. O. Walther, Vol. **5**, Springer-Verlag, 1996, pp. 99-189. MR1393487 (97k:58116)

[73] X. Lin, J. So and J. Wu, Center manifolds for partial differential equations with delays, *Proc. Roy. Soc. Edinburgh* **122A** (1992), 237-254. MR1200199 (93j:34116)

[74] Z. Liu, P. Magal and S. Ruan, Projectors on the generalized eigenspaces for functional differential equations using integrated semigroup, *J. Differential Equations* **244** (2008), 1784–1809. MR2404439 (2009b:34191)

[75] P. Magal, Compact attractors for time periodic age-structured population models, *Electr. J. Differential Equations* **2001** (2001), No. 65, 1-35. MR1863784 (2002k:37160)

[76] P. Magal and S. Ruan, On integrated semigroups and age-structured models in L^p space, *Differential Integral Equations* **20** (2007), 197-239. MR2294465 (2008c:47066)

[77] P. Magal and S. Ruan (eds.), *Structured Population Models in Biology and Epidemiology*, Lect. Notes Math. Vol. **1936**, Springer-Verlag, Berlin, 2008. MR2445337

[78] P. Magal and S. Ruan, On semilinear Cauchy problems with non-dense domain, *Adv. Diff. Equations* (in press).

[79] P. Magal and S. Ruan, Sustained oscillations in an evolutionary epidemiological model of influenza A drift (submitted).

[80] P. Magal and H. R. Thieme, Eventual compactness for semiflows generated by nonlinear age-structured models, *Comm. Pure Appl. Anal.* **3** (2004), 695-727. MR2106296 (2005h:34161)

[81] A. Mielke, A reduction principle for nonautonomous systems in infinite-dimensional spaces, *J. Differential Equations* **65** (1986), 68-88. MR859473 (87k:47140)

[82] A. Mielke, Normal hyperbolicity of center manifolds and Saint-Vernant's principle, *Arch. Rational Mech. Anal.* **110** (1990), 353-372. MR1049211 (91e:58171)

[83] Nguyen Van Minh and J. Wu, Invariant manifolds of partial functional differential equations, *J. Differential Equations* **198** (2004), 381-421. MR2039148 (2005a:34102)

[84] F. Neubrander, Integrated semigroups and their application to the abstract Cauchy problem, *Pac. J. Math.* **135** (1988), 111-155. MR965688 (90b:47073)

[85] A. Pazy, *Semigroups of Linear Operator and Applications to Partial Differential Equations*, Springer-Verlag, New York, 1983. MR710486 (85g:47061)

[86] C. M. Pease, An evolutionary epidemiological mechanism with application to type A influenza, *Theoret. Pop. Biol.* **31** (1987), 422-452.

[87] O. Perron, Über stabilität und asymptotische verhalten der integrale von differentialgleichungssystemen, *Math. Z.* **29** (1928), 129-160. MR1544998

[88] V. A. Pliss, Principal reduction in the theory of stability of motion, *Izv. Akad. Nauk. SSSR Mat. Ser.* **28** (1964), 1297-1324. MR0190449 (32:7861)

[89] J. Prüss, On the qualitative behavior of populations with age-speciific interactions, *Comput. Math. Appl.* **9** (1983), 327-339. MR702651 (84h:92035)

[90] B. Sandstede, Center manifolds for homoclinic solutions, *J. Dynam. Differential Equations* **12** (2000), 449-510. MR1800130 (2001m:37167)

[91] B. Scarpellini, Center manifolds of infinite dimensions I: Main results and applications, *ZAMP* **42** (1991), 1-32. MR1102229 (92i:58170)

[92] H. H. Schaefer, *Banach Lattice and Positive Operator*, Springer-Verlag, Berlin, 1974. MR0423039 (54:11023)

[93] A. Scheel, Radially symmetric patterns of reaction-diffusion systems, *Mem. Amer. Math. Soc.* **165** (2003), No. 786. MR1997690 (2005c:35160)

[94] G. R. Sell and Y. You, *Dynamics of Evolutionary Equations*, Springer-Verlag, New York, 2002. MR1873467 (2003f:37001b)

[95] J. Sijbrand, Properties of center manifolds, *Trans. Amer. Math. Soc.* **289** (1985), 431-469. MR783998 (86i:58099)

[96] J. H. Swart, Hopf bifurcation and the stability of non-linear age-depedent population models, *Comput. Math. Appl.* **15** (1988), 555-564. MR953565 (89g:92052)

[97] A. E. Taylor and D. C. Lay, *Introduction to Functional Analysis*, John Wiley & Sons, New York, 1980. MR564653 (81b:46001)

[98] H. R. Thieme, Semiflows generated by Lipschitz perturbations of non-densely defined operators, *Differential Integral Equations* **3** (1990), 1035-1066. MR1073056 (92e:47121)

[99] H. R. Thieme, "Integrated semigroups" and integrated solutions to abstract Cauchy problems, *J. Math. Anal. Appl.* **152** (1990), 416-447. MR1077937 (91k:47093)

[100] H. R. Thieme, Stability change for the endemic equilibrium in age-structured models for the spread of S-I-R type infectious diseases, in *"Differential Equation Models in Biology,*

Epidemiology and Ecology", eds. by S. N. Busenberg and M. Martelli, Lect. Notes in Biomath. **92**, Springer, Berlin, 1991, pp. 139-158. MR1193478 (93h:92034)

[101] H. R. Thieme, Quasi-compact semigroups via bounded perturbation, in *"Advances in Mathematical Population Dynamics-Molecules, Cells and Man"*, eds. by O. Arino, D. Axelrod and M. Kimmel, World Sci. Publ., River Edge, NJ, 1997, pp. 691-713. MR1634223 (99i:47070)

[102] H. R. Thieme, Positive perturbation of operator semigroups: Growth bounds, essential compactness, and asynchronous exponential growth, *Discrete Contin. Dynam. Systems* **4** (1998), 735-764. MR1641201 (2000e:47069)

[103] A. Vanderbauwhede, Invariant manifolds in infinite dimensions, in *"Dynamics of Infinite Dimensional Systems"* , ed. by S. N. Chow and J. K. Hale, Springer-Verlag, Berlin, 1987, pp. 409-420. MR921925 (89e:47098)

[104] A. Vanderbauwhede, Center manifold, normal forms and elementary bifurcations, *Dynamics Reported*, ed. by U. Kirchgraber and H. O. Walther, Vol. **2**, John Wiley & Sons, 1989, pp. 89-169. MR1000977 (90g:58092)

[105] A. Vanderbauwhede and S. A. van Gils, Center manifolds and contractions on a scale of Banach spaces, *J. Funct. Anal.* **72** (1987), 209-224. MR886811 (88d:58085)

[106] A. Vanderbauwhede and G. Iooss, Center manifold theory in infinite dimensions, *Dynamics Reported* (new series), ed. by C. K. R. T. Jones, U. Kirchgraber and H. O. Walther, Vol. **1**, Springer-Verlag, Berlin, 1992, pp. 125-163. MR1153030 (93f:58174)

[107] G. F. Webb, An age-dependent epidenuc model with spatial diffusion, *Arch. Rational Mech. Anal.* **75** (1980), 91-102. MR592106 (81k:92052)

[108] G. F. Webb, *Theory of Nonlinear Age-Dependent Population Dynamics*, Marcel Dekker, New York, 1985. MR772205 (86e:92032)

[109] G. F. Webb, E. M. C. D'Agata, P. Magal and S. Ruan, A model of antibiotic resistant bacterial epidemics in hospitals, *Proc. Natl. Acad. Sci. USA* **102** (2005), 13343-13348.

[110] S. Wiggins, *Normally Hyperbolic Invariant Manifolds in Dynamical Systems*, Springer-Verlag, New York, 1994. MR1278264 (95g:58163)

[111] J. Wu, *Theory and Applications of Partial Differential Equations*, Springer-Verlag, New York, 1996. MR1415838 (98a:35135)

[112] Y. Yi, A generalized integral manifold theorem, *J. Differential Equations* **102** (1993), 153-187. MR1209981 (94c:58148)

[113] K. Yosida, *Functional Analysis*, Springer-Verlag, Berlin, 1980. MR617913 (82i:46002)

[114] P. Zhang, Z. Feng and F. Milner, A schistosomiasis model with an age-structure in human hosts and its applicationo to treatment strategies, *Math. Biosci.* **205** (2007), 83-107. MR2290375 (2007i:92062)

Editorial Information

To be published in the *Memoirs*, a paper must be correct, new, nontrivial, and significant. Further, it must be well written and of interest to a substantial number of mathematicians. Piecemeal results, such as an inconclusive step toward an unproved major theorem or a minor variation on a known result, are in general not acceptable for publication.

Papers appearing in *Memoirs* are generally at least 80 and not more than 200 published pages in length. Papers less than 80 or more than 200 published pages require the approval of the Managing Editor of the Transactions/Memoirs Editorial Board. Published pages are the same size as those generated in the style files provided for $\mathcal{AMS}$-LaTeX or $\mathcal{AMS}$-TeX.

Information on the backlog for this journal can be found on the AMS website starting from http://www.ams.org/memo.

A Consent to Publish and Copyright Agreement is required before a paper will be published in the *Memoirs*. After a paper is accepted for publication, the Providence office will send a Consent to Publish and Copyright Agreement to all authors of the paper. By submitting a paper to the *Memoirs*, authors certify that the results have not been submitted to nor are they under consideration for publication by another journal, conference proceedings, or similar publication.

Information for Authors

Memoirs is an author-prepared publication. Once formatted for print and on-line publication, articles will be published as is with the addition of AMS-prepared frontmatter and backmatter. Articles are not copyedited; however, confirmation copy will be sent to the authors.

Initial submission. The AMS uses Centralized Manuscript Processing for initial submissions. Authors should submit a PDF file using the Initial Manuscript Submission form found at www.ams.org/peer-review-submission, or send one copy of the manuscript to the following address: Centralized Manuscript Processing, MEMOIRS OF THE AMS, 201 Charles Street, Providence, RI 02904-2294 USA. If a paper copy is being forwarded to the AMS, indicate that it is for *Memoirs* and include the name of the corresponding author, contact information such as email address or mailing address, and the name of an appropriate Editor to review the paper (see the list of Editors below).

The paper must contain a *descriptive title* and an *abstract* that summarizes the article in language suitable for workers in the general field (algebra, analysis, etc.). The *descriptive title* should be short, but informative; useless or vague phrases such as "some remarks about" or "concerning" should be avoided. The *abstract* should be at least one complete sentence, and at most 300 words. Included with the footnotes to the paper should be the 2010 *Mathematics Subject Classification* representing the primary and secondary subjects of the article. The classifications are accessible from www.ams.org/msc/. The Mathematics Subject Classification footnote may be followed by a list of *key words and phrases* describing the subject matter of the article and taken from it. Journal abbreviations used in bibliographies are listed in the latest *Mathematical Reviews* annual index. The series abbreviations are also accessible from www.ams.org/msnhtml/serials.pdf. To help in preparing and verifying references, the AMS offers MR Lookup, a Reference Tool for Linking, at www.ams.org/mrlookup/.

Electronically prepared manuscripts. The AMS encourages electronically prepared manuscripts, with a strong preference for $\mathcal{AMS}$-LaTeX. To this end, the Society has prepared $\mathcal{AMS}$-LaTeX author packages for each AMS publication. Author packages include instructions for preparing electronic manuscripts, samples, and a style file that generates the particular design specifications of that publication series. Though $\mathcal{AMS}$-LaTeX is the highly preferred format of TeX, author packages are also available in $\mathcal{AMS}$-TeX.

Authors may retrieve an author package for *Memoirs of the AMS* from www.ams.org/journals/memo/memoauthorpac.html or via FTP to ftp.ams.org (login as anonymous, enter your complete email address as password, and type cd pub/author-info). The

AMS Author Handbook and the *Instruction Manual* are available in PDF format from the author package link. The author package can also be obtained free of charge by sending email to `tech-support@ams.org` (Internet) or from the Publication Division, American Mathematical Society, 201 Charles St., Providence, RI 02904-2294, USA. When requesting an author package, please specify $\mathcal{AMS}$-LaTeX or $\mathcal{AMS}$-TeX and the publication in which your paper will appear. Please be sure to include your complete mailing address.

After acceptance. The source files for the final version of the electronic manuscript should be sent to the Providence office immediately after the paper has been accepted for publication. The author should also submit a PDF of the final version of the paper to the editor, who will forward a copy to the Providence office.

Accepted electronically prepared files can be submitted via the web at `www.ams.org/submit-book-journal/`, sent via FTP, or sent on CD-Rom or diskette to the Electronic Prepress Department, American Mathematical Society, 201 Charles Street, Providence, RI 02904-2294 USA. TeX source files and graphic files can be transferred over the Internet by FTP to the Internet node `ftp.ams.org` (130.44.1.100). When sending a manuscript electronically via CD-Rom or diskette, please be sure to include a message indicating that the paper is for the *Memoirs*.

Electronic graphics. Comprehensive instructions on preparing graphics are available at `www.ams.org/authors/journals.html`. A few of the major requirements are given here.

Submit files for graphics as EPS (Encapsulated PostScript) files. This includes graphics originated via a graphics application as well as scanned photographs or other computer-generated images. If this is not possible, TIFF files are acceptable as long as they can be opened in Adobe Photoshop or Illustrator.

Authors using graphics packages for the creation of electronic art should also avoid the use of any lines thinner than 0.5 points in width. Many graphics packages allow the user to specify a "hairline" for a very thin line. Hairlines often look acceptable when proofed on a typical laser printer. However, when produced on a high-resolution laser imagesetter, hairlines become nearly invisible and will be lost entirely in the final printing process.

Screens should be set to values between 15% and 85%. Screens which fall outside of this range are too light or too dark to print correctly. Variations of screens within a graphic should be no less than 10%.

Inquiries. Any inquiries concerning a paper that has been accepted for publication should be sent to `memo-query@ams.org` or directly to the Electronic Prepress Department, American Mathematical Society, 201 Charles St., Providence, RI 02904-2294 USA.

Editors

This journal is designed particularly for long research papers, normally at least 80 pages in length, and groups of cognate papers in pure and applied mathematics. Papers intended for publication in the *Memoirs* should be addressed to one of the following editors. The AMS uses Centralized Manuscript Processing for initial submissions to AMS journals. Authors should follow instructions listed on the Initial Submission page found at www.ams.org/memo/memosubmit.html.

Algebra, to ALEXANDER KLESHCHEV, Department of Mathematics, University of Oregon, Eugene, OR 97403-1222; e-mail: ams@noether.uoregon.edu

Algebraic geometry, to DAN ABRAMOVICH, Department of Mathematics, Brown University, Box 1917, Providence, RI 02912; e-mail: amsedit@math.brown.edu

Algebraic geometry and its applications, to MINA TEICHER, Emmy Noether Research Institute for Mathematics, Bar-Ilan University, Ramat-Gan 52900, Israel; e-mail: teicher@macs.biu.ac.il

Algebraic topology, to ALEJANDRO ADEM, Department of Mathematics, University of British Columbia, Room 121, 1984 Mathematics Road, Vancouver, British Columbia, Canada V6T 1Z2; e-mail: adem@math.ubc.ca

Combinatorics, to JOHN R. STEMBRIDGE, Department of Mathematics, University of Michigan, Ann Arbor, Michigan 48109-1109; e-mail: JRS@umich.edu

Commutative and homological algebra, to LUCHEZAR L. AVRAMOV, Department of Mathematics, University of Nebraska, Lincoln, NE 68588-0130; e-mail: avramov@math.unl.edu

Complex analysis and harmonic analysis, to ALEXANDER NAGEL, Department of Mathematics, University of Wisconsin, 480 Lincoln Drive, Madison, WI 53706-1313; e-mail: nagel@math.wisc.edu

Differential geometry and global analysis, to CHRIS WOODWARD, Department of Mathematics, Rutgers University, 110 Frelinghuysen Road, Piscataway, NJ 08854; e-mail: ctw@math.rutgers.edu

Dynamical systems and ergodic theory and complex analysis, to YUNPING JIANG, Department of Mathematics, CUNY Queens College and Graduate Center, 65-30 Kissena Blvd., Flushing, NY 11367; e-mail: Yunping.Jiang@qc.cuny.edu

Functional analysis and operator algebras, to DIMITRI SHLYAKHTENKO, Department of Mathematics, University of California, Los Angeles, CA 90095; e-mail: shlyakht@math.ucla.edu

Geometric analysis, to WILLIAM P. MINICOZZI II, Department of Mathematics, Johns Hopkins University, 3400 N. Charles St., Baltimore, MD 21218; e-mail: trans@math.jhu.edu

Geometric topology, to MARK FEIGHN, Math Department, Rutgers University, Newark, NJ 07102; e-mail: feighn@andromeda.rutgers.edu

Harmonic analysis, representation theory, and Lie theory, to ROBERT J. STANTON, Department of Mathematics, The Ohio State University, 231 West 18th Avenue, Columbus, OH 43210-1174; e-mail: stanton@math.ohio-state.edu

Logic, to STEFFEN LEMPP, Department of Mathematics, University of Wisconsin, 480 Lincoln Drive, Madison, Wisconsin 53706-1388; e-mail: lempp@math.wisc.edu

Number theory, to JONATHAN ROGAWSKI, Department of Mathematics, University of California, Los Angeles, CA 90095; e-mail: jonr@math.ucla.edu

Number theory, to SHANKAR SEN, Department of Mathematics, 505 Malott Hall, Cornell University, Ithaca, NY 14853; e-mail: ss70@cornell.edu

Partial differential equations, to GUSTAVO PONCE, Department of Mathematics, South Hall, Room 6607, University of California, Santa Barbara, CA 93106; e-mail: ponce@math.ucsb.edu

Partial differential equations and dynamical systems, to PETER POLACIK, School of Mathematics, University of Minnesota, Minneapolis, MN 55455; e-mail: polacik@math.umn.edu

Probability and statistics, to RICHARD BASS, Department of Mathematics, University of Connecticut, Storrs, CT 06269-3009; e-mail: bass@math.uconn.edu

Real analysis and partial differential equations, to DANIEL TATARU, Department of Mathematics, University of California, Berkeley, Berkeley, CA 94720; e-mail: tataru@math.berkeley.edu

All other communications to the editors, should be addressed to the Managing Editor, ROBERT GURALNICK, Department of Mathematics, University of Southern California, Los Angeles, CA 90089-1113; e-mail: guralnic@math.usc.edu.

Titles in This Series

951 **Pierre Magal and Shigui Ruan,** Center manifolds for semilinear equations with non-dense domain and applications to Hopf bifurcation in age structured models, 2009

950 **Cédric Villani,** Hypocoercivity, 2009

949 **Drew Armstrong,** Generalized noncrossing partitions and combinatorics of Coxeter groups, 2009

948 **Nan-Kuo Ho and Chiu-Chu Melissa Liu,** Yang-Mills connections on orientable and nonorientable surfaces, 2009

947 **W. Turner,** Rock blocks, 2009

946 **Jay Jorgenson and Serge Lang,** Heat Eisenstein series on $SL_n(C)$, 2009

945 **Tobias H. Jäger,** The creation of strange non-chaotic attractors in non-smooth saddle-node bifurcations, 2009

944 **Yuri Kifer,** Large deviations and adiabatic transitions for dynamical systems and Markov processes in fully coupled averaging, 2009

943 **István Berkes and Michel Weber,** On the convergence of $\sum c_k f(n_k x)$, 2009

942 **Dirk Kussin,** Noncommutative curves of genus zero: Related to finite dimensional algebras, 2009

941 **Gelu Popescu,** Unitary invariants in multivariable operator theory, 2009

940 **Gérard Iooss and Pavel I. Plotnikov,** Small divisor problem in the theory of three-dimensional water gravity waves, 2009

939 **I. D. Suprunenko,** The minimal polynomials of unipotent elements in irreducible representations of the classical groups in odd characteristic, 2009

938 **Antonino Morassi and Edi Rosset,** Uniqueness and stability in determining a rigid inclusion in an elastic body, 2009

937 **Skip Garibaldi,** Cohomological invariants: Exceptional groups and spin groups, 2009

936 **André Martinez and Vania Sordoni,** Twisted pseudodifferential calculus and application to the quantum evolution of molecules, 2009

935 **Mihai Ciucu,** The scaling limit of the correlation of holes on the triangular lattice with periodic boundary conditions, 2009

934 **Arjen Doelman, Björn Sandstede, Arnd Scheel, and Guido Schneider,** The dynamics of modulated wave trains, 2009

933 **Luchezar Stoyanov,** Scattering resonances for several small convex bodies and the Lax-Phillips conjuecture, 2009

932 **Jun Kigami,** Volume doubling measures and heat kernel estimates of self-similar sets, 2009

931 **Robert C. Dalang and Marta Sanz-Solé,** Hölder-Sobolv regularity of the solution to the stochastic wave equation in dimension three, 2009

930 **Volkmar Liebscher,** Random sets and invariants for (type II) continuous tensor product systems of Hilbert spaces, 2009

929 **Richard F. Bass, Xia Chen, and Jay Rosen,** Moderate deviations for the range of planar random walks, 2009

928 **Ulrich Bunke,** Index theory, eta forms, and Deligne cohomology, 2009

927 **N. Chernov and D. Dolgopyat,** Brownian Brownian motion-I, 2009

926 **Riccardo Benedetti and Francesco Bonsante,** Canonical wick rotations in 3-dimensional gravity, 2009

925 **Sergey Zelik and Alexander Mielke,** Multi-pulse evolution and space-time chaos in dissipative systems, 2009

924 **Pierre-Emmanuel Caprace,** "Abstract" homomorphisms of split Kac-Moody groups, 2009

923 **Michael Jöllenbeck and Volkmar Welker,** Minimal resolutions via algebraic discrete Morse theory, 2009

922 **Ph. Barbe and W. P. McCormick,** Asymptotic expansions for infinite weighted convolutions of heavy tail distributions and applications, 2009

TITLES IN THIS SERIES

921 **Thomas Lehmkuhl,** Compactification of the Drinfeld modular surfaces, 2009
920 **Georgia Benkart, Thomas Gregory, and Alexander Premet,** The recognition theorem for graded Lie algebras in prime characteristic, 2009
919 **Roelof W. Bruggeman and Roberto J. Miatello,** Sum formula for SL_2 over a totally real number field, 2009
918 **Jonathan Brundan and Alexander Kleshchev,** Representations of shifted Yangians and finite W-algebras, 2008
917 **Salah-Eldin A. Mohammed, Tusheng Zhang, and Huaizhong Zhao,** The stable manifold theorem for semilinear stochastic evolution equations and stochastic partial differential equations, 2008
916 **Yoshikata Kida,** The mapping class group from the viewpoint of measure equivalence theory, 2008
915 **Sergiu Aizicovici, Nikolaos S. Papageorgiou, and Vasile Staicu,** Degree theory for operators of monotone type and nonlinear elliptic equations with inequality constraints, 2008
914 **E. Shargorodsky and J. F. Toland,** Bernoulli free-boundary problems, 2008
913 **Ethan Akin, Joseph Auslander, and Eli Glasner,** The topological dynamics of Ellis actions, 2008
912 **Igor Chueshov and Irena Lasiecka,** Long-time behavior of second order evolution equations with nonlinear damping, 2008
911 **John Locker,** Eigenvalues and completeness for regular and simply irregular two-point differential operators, 2008
910 **Joel Friedman,** A proof of Alon's second eigenvalue conjecture and related problems, 2008
909 **Cameron McA. Gordon and Ying-Qing Wu,** Toroidal Dehn fillings on hyperbolic 3-manifolds, 2008
908 **J.-L. Waldspurger,** L'endoscopie tordue n'est pas si tordue, 2008
907 **Yuanhua Wang and Fei Xu,** Spinor genera in characteristic 2, 2008
906 **Raphaël S. Ponge,** Heisenberg calculus and spectral theory of hypoelliptic operators on Heisenberg manifolds, 2008
905 **Dominic Verity,** Complicial sets characterising the simplicial nerves of strict ω-categories, 2008
904 **William M. Goldman and Eugene Z. Xia,** Rank one Higgs bundles and representations of fundamental groups of Riemann surfaces, 2008
903 **Gail Letzter,** Invariant differential operators for quantum symmetric spaces, 2008
902 **Bertrand Toën and Gabriele Vezzosi,** Homotopical algebraic geometry II: Geometric stacks and applications, 2008
901 **Ron Donagi and Tony Pantev (with an appendix by Dmitry Arinkin),** Torus fibrations, gerbes, and duality, 2008
900 **Wolfgang Bertram,** Differential geometry, Lie groups and symmetric spaces over general base fields and rings, 2008
899 **Piotr Hajłasz, Tadeusz Iwaniec, Jan Malý, and Jani Onninen,** Weakly differentiable mappings between manifolds, 2008
898 **John Rognes,** Galois extensions of structured ring spectra/Stably dualizable groups, 2008
897 **Michael I. Ganzburg,** Limit theorems of polynomial approximation with exponential weights, 2008

For a complete list of titles in this series, visit the
AMS Bookstore at **www.ams.org/bookstore/**.